I0479383

ASTROPHYSICAL COLLECTION

THE STARS OF THE UNIVERSE

volume 1

JOSE RUIZ WATZECK

SUMMARY

SUMMARY

The stars are one of the most fascinating entities in the universe, and since ancient times they have been the subject of study and wonder. With the advent of modern technology, we were able to better discover and understand the nature of these cosmic entities, which are the building blocks of the universe.

In this book, we will explore the largest known stars in the universe, which have unimaginable dimensions and challenge our understanding of stellar physics. Varying in size, brightness, and age, these stars offer unique insights into the evolution and dynamics of the universe.

The formation of a giant star begins with the gravitational collapse of a molecular cloud of gas and dust. As the cloud contracts, the temperature and density at its core increase until nuclear ignition occurs, initiating the fusion of hydrogen into helium. The energy released by this process sustains the star, which enters a hydrostatic balance between the force of gravity and radiation pressure.

However, the largest stars in the universe follow a different evolutionary path. Since they have a much greater mass than the Sun, they burn up their nuclear fuel much faster. As a result, their lifespan is significantly shorter, and their final fate is very different.

As the star nears the end of its life, it undergoes a series of thermonuclear explosions that culminate in a supernova. This releases an incredible amount of energy and can lead to the formation of compact stellar objects, such as black holes or neutron stars.

The internal structure of a giant star is influenced by its mass, temperature, and age. As the star ages, it expands and cools, resulting in an increasingly thin atmosphere and an increasingly dense core.

Giant stars are known for their high luminosity, which is a measure of the amount of energy they emit. This is because these stars have a very high rate of nuclear fusion at their core, which results in the release of enormous amounts of energy in the form of electromagnetic radiation. Some of these stars can emit more than a million times the luminosity of the Sun.

Giant stars have significant implications for the evolution of the universe, they are responsible for the production of heavy elements, such as iron, which are essential for planet formation and life. Also, a supernova explosion can result in the formation of new stars and planetary systems.

However, giant stars can also pose a danger to life in the universe, a supernova explosion can be extremely destructive and can wipe out all life in a nearby star system.

Astronomical measurements are used to study celestial objects and understand the universe. These measurements are made using special units to quantify distances, sizes, masses, and other properties of celestial bodies.

Some of the more common units used in astronomy include: Astronomical unit (AU): used to measure distances within the solar system, corresponding to the average distance between the Earth and the Sun, about 150 million kilometers.

Light year (AL): used to measure distances outside the solar system, corresponding to the distance that light travels in one year, equivalent to 9.5 trillion kilometers.

Parsec (pc) – Another unit of distance measurement outside the solar system, corresponding to the distance at which a star would have a parallax of one arcsecond, representing 3.2 AL (light year). We can also apply the measurement of megaparsecs and gigaparsecs to greater distances, however, a topic for a future book.

Apparent Magnitude – Used to measure the brightness of celestial objects, with smaller numbers indicating greater brightness.
Absolute Magnitude: It is used to measure the intrinsic luminosity of a celestial object, adjusting its apparent magnitude based on its distance.

Radian (rad): used to measure angles in the sky, corresponding to the central angle subtended by an arc of length equal to the radius of the circumference.
These astronomical measurements are essential for the investigation and understanding of the universe, and are used in several areas of astronomy, such as astrophysics, astrobiology, and cosmology.

To conclude, the stars are true cosmic colossi that challenge our understanding of the universe. Its size, brightness, and evolution present a unique set of challenges to stellar physics and our understanding of the dynamics of the universe. Furthermore, these stars have significant implications for the evolution of the universe and could play a crucial role in the formation of planets and life. This book offers a detailed and accessible look at these extraordinary celestial phenomena and their importance to our understanding of the universe.

THE SUN

In relation to all the bodies in our solar system, such as comets, stardust, asteroids, planets, natural satellites, etc., orbit this star. Classified as a yellow dwarf,responsible for 99.86% of thepastaof the Solar System, the Sun has a mass 332,900 times that of the Earth.Land, it's yoursvolumeIt is 1.3 million times greater than that of our planet. The distance from the Earth to the Sun is about 150 millionkilometresor 1astronomical unit(AU). This distance varies throughout the year, from a minimum of 147.1 million kilometers (0.9833 AU) at perihelion[1], to a maximum of 152.1 million kilometers (1.017 AU), inaphelion[2] (which occurs around the dayJuly 4th).

Light from the sun takes about 500 seconds, or 8 minutes and 34 seconds to reach Earth, its primary composition is 74% of its mass or 91% of its volume, it constitutes hydrogen, 24% of its mass or the 7% of its volume, is constituted by helium and the other elements being around 2% of its volume, constitutes in; calcium, chromium, sulfur, iron, neon, nickel, oxygen and silicon. Its spectral class is known as G2V,its temperature varies according to the layer of its structure. The core, which corresponds to the central portion of the solar structure, is also its hottest region. It is in it that the process of fusion of hydrogen atoms occurs, resulting in the formation of helium. Nuclear fusion is responsible for generating heat that spreads to other layers. Thus, the temperature of the core of the Sun reaches 15.7 million degrees Celsius. At the solar surface, which is called the photosphere, the temperature is much lower than at the core, reaching 5,500 °C. The convective zone, which consists of an intermediate layer, has temperatures of up to two million degrees Celsius or5780 degrees Kelvin[3]or 5,780K where its original color is white, although here on Earth it is seen in yellow, orange and sometimes reddish when it is on the horizon.The origin of the Sun is associated with the gravitational collapse of the solar nebula, a cloud formed by dust and gases, this process began about 4,500 million years ago, which corresponds to the age of the Sun.

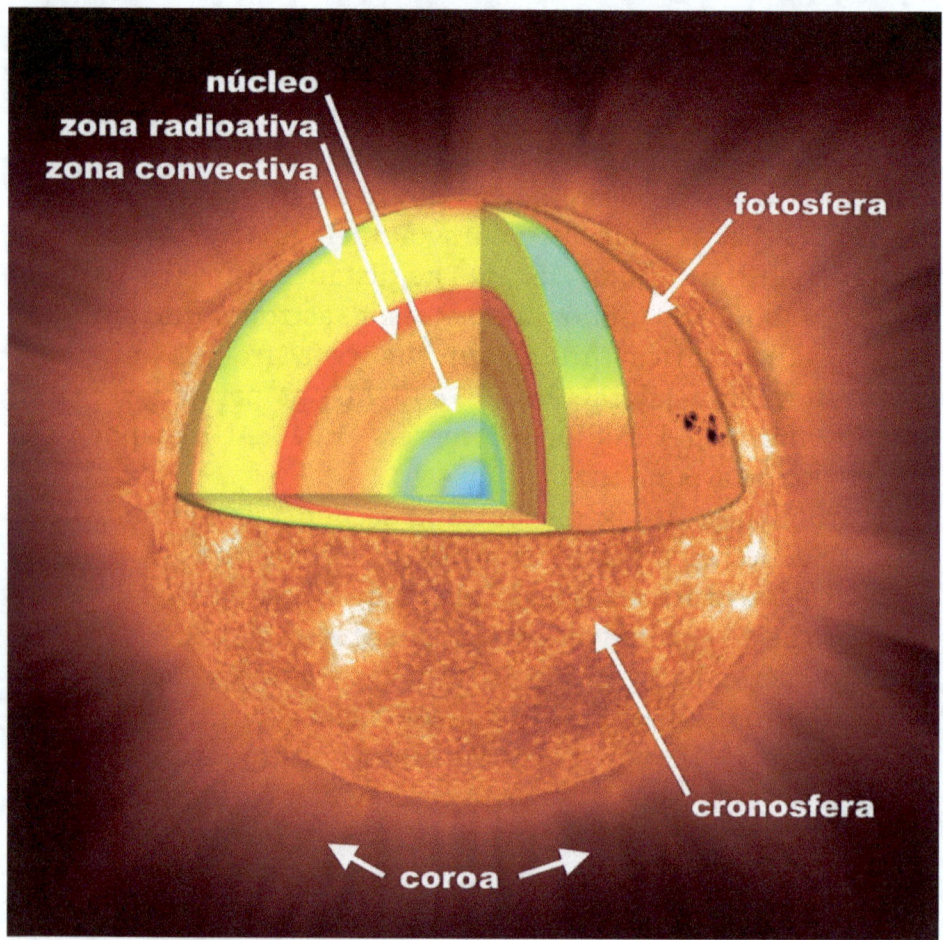

Diagram indicating each of the six layers that make up the Sun.

• **Center:**It corresponds to the innermost layer of the Sun. It is approximately a thousand times the size of the Earth and is also denser than our planet. As we saw earlier, it is in the core of the Sun where the nuclear reactions responsible for the production of helium atoms take place. As a result of this process, the emission of light and the generation of heat take place.

- **Radiative zone:**it is an extensive layer that surrounds the core, corresponding to almost half the radius of the Sun. The energy that is generated in the solar core is radiated through this region, where the temperature drops significantly in comparison with the first layer.

- **convective zone:**Also called the convection zone, it corresponds to the layer located above the radiative zone. In it, energy is transferred by means of convection currents formed by the movement of gases at high temperatures.

- **Photosphere:**corresponds to the surface of the Sun. With the help of appropriate instruments it is possible to observe the thermal columns that rise from the convective zone towards the photosphere, which appear in the form of granules. Dark spots are also observed and are called sunspots.

- **Atmosphere:**forms the solar atmosphere, just above the photosphere. It has a pink color and lower temperatures, around 4,700 °C. Gas jets are emitted from this layer toward the corona.

- **Crown:**outermost layer of the sun's atmosphere. The corona is much hotter than the layers below it, reaching 2 million degrees Celsius in the areas furthest from the surface. It consists of a very extensive region, millions of kilometers long, made up of gases in motion. Its speed is variable and can reach 400 km/s. This is where the solar wind forms.

There is no solid surface on the Sun, and for this reason it is difficult to determine how many days it takes to complete one rotation. It is estimated that, at its equatorial line, this movement takes 25 Earth days, and at the poles it takes longer, 36 Earth days.

The life cycle of the sun

stellar evolutionis measured in two ways: through the current age ofsequence, which is determined throughcomputational modelingof stellar evolution; Isnucleocosmochronology[4]. The age measured using these procedures is in agreement with theradiometric age[5]of the oldest material found in the Solar System, which is 4,567 million years old.

The Sun is about halfway through the main sequence, the period during which nuclear fusion fuses hydrogen into helium. Every second, more than 4 million tons of matter are converted into energy inside the solar center, producing neutrinos and solar radiation. At this rate, the Sun has converted about 100 Earth masses into energy from its formation to the present. The Sun will remain on the main sequence for about 10 billion (10 billion) years.In about 5 billion years, the hydrogen in the solar core will run out. When this happens, the Sun will contract under its own gravity, raising the temperature of the solar core to 100 million kelvins, enough to initiate thehelium nuclear fusion, producingcoal, entering the phase ofasymptotic giant branch.

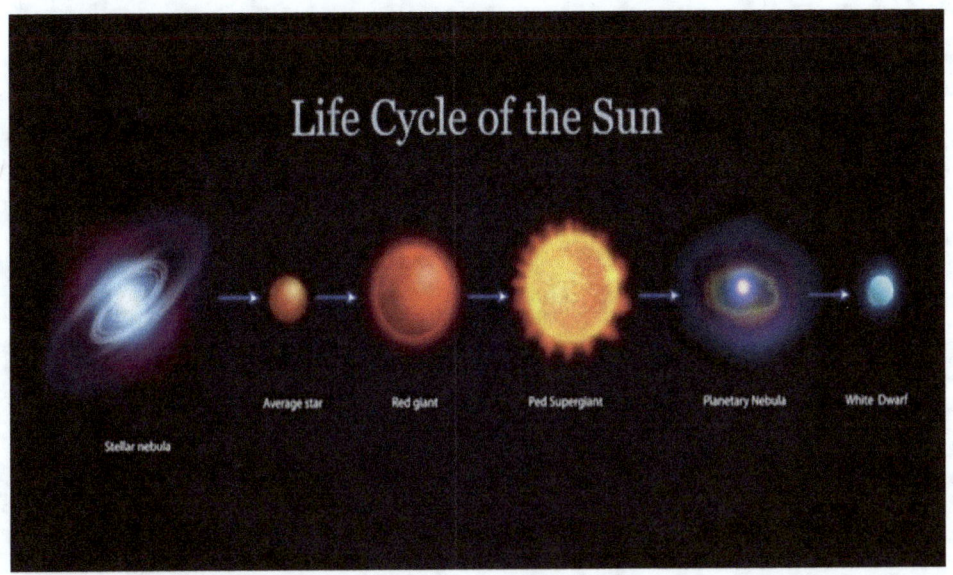

solar energy production

Hydrogen fusion occurs primarily in a chain of reactions calledproton-proton chain:

$$4\,^1H \rightarrow 2\,^2H + 2\,y^+ + 2v_{Is}\,(4.0\text{ MeV} + 1.0\text{ MeV})$$

$$2\,^1H + 2\,^2H \rightarrow 2\,^3He + 2\gamma\,(5.5\text{ MeV})$$

$$\text{two}^3\text{the} \rightarrow\,^4He + 2\,^1H\,(12.9\text{ MeV})$$

These reactions can be summarized according to the following formula:

$$4\,^1H \rightarrow\,^4\text{the} + 2\text{ and}^+ + 2v_{Is} + 2\,\gamma\,(26.7\text{ MeV})$$

The Sun has about 8.9 x 1056 hydrogen nuclei (free protons), and the proton-proton chain occurs 9.2 x 1037 times per second in the solar core. Since this reaction uses four protons, about 3.7 x 1038 protons (or 6.2 x 1011 kg) are converted to helium nuclei every second.[This reaction converts 0.7% of the melt into energy, and as a result, about 4.26 million metric tons per second is converted into 383 yotta-watts (3.83 x 1026 W), or 9.15 x 1010 megatons ofT.N.T.of energy per second, according to the mass-energy equationE=mc²inAlbert Einstein.

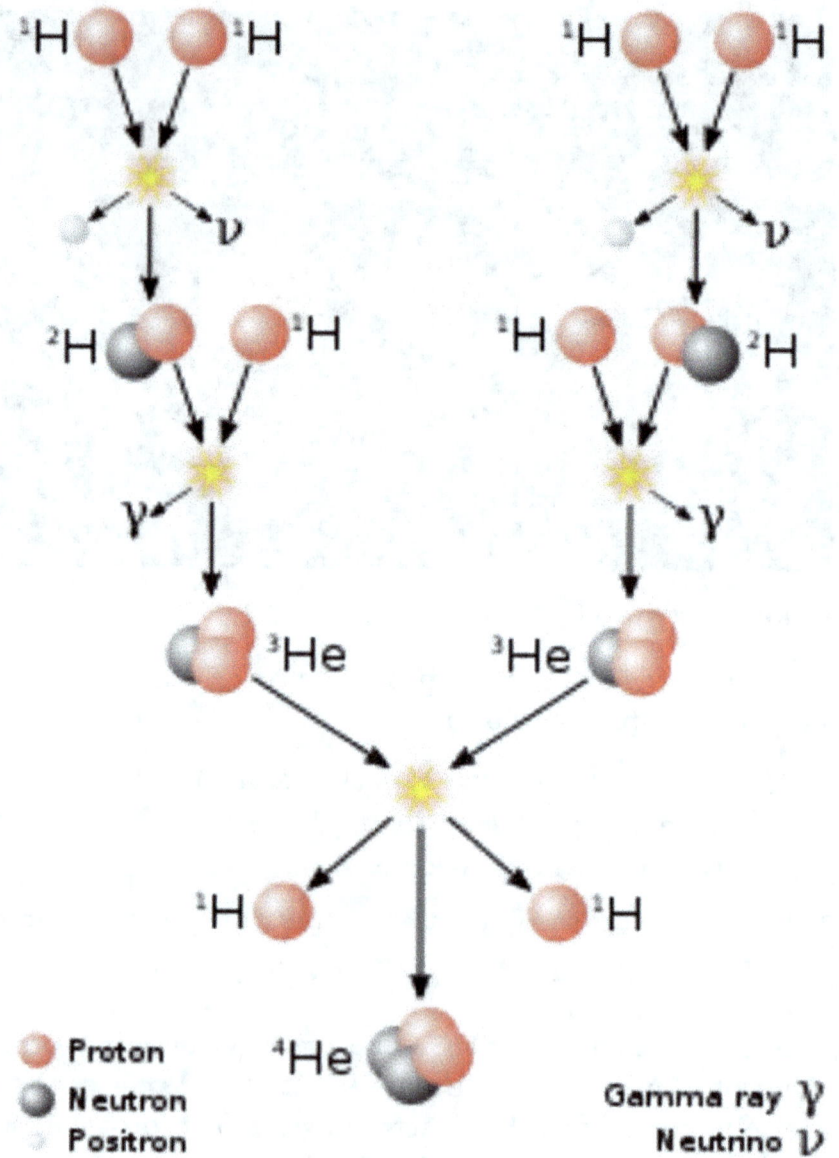

diagram ofproton-proton chain, the cycle ofNuclear
fusiongenerates most of the sun's energy

The power density is about 194 μW/kg of matter, and although fusion takes place in the relatively small solar core, the plasma power density in this region is 150 times higher. By comparison, the heat produced by the human body is 1.3 W/kg, about 600 times that of the Sun, per unit mass.

Even taking into account only the solar core, with densities 150 times greater than the average density of the star, the Sun produces relatively little energy, at a rate of 0.272 W/m³. Surprisingly, this power is much less than that generated by a burning candle. Using plasma on Earth with parameters similar to those in the solar core is impossible, even a modest 1 GW plant would require around 5 billion (5 billion) metric tons of plasma.

The rate of nuclear fusion is highly dependent on the density and temperature of the core: a slightly higher fusion rate causes the core to heat up, expanding the outer layers of the Sun and consequently decreasing the gravitational pressure exerted by the outer layers. . and the fusion rate. As the melt rate decreases, the outer layers contract, increasing their pressure against the solar core, which will again increase the melt rate, causing the cycle to repeat itself.

The high-energy photons (gamma rays) generated by nuclear fusion are absorbed by the nuclei present in the solar plasma and re-emitted again in a random direction, this time with slightly lower energy. They are then absorbed again and the cycle repeats. As a result, radiation generated by nuclear fusion in the solar core takes a long time to reach the surface. Travel time estimates range from 10 to 170,000 years.

After passing through the convection layer to the "transparent" surface of the photosphere, the photons escape as visible light. Each gamma ray from the solar core is converted to several million visible photons before escaping into space. Neutrinos are also generated by nuclear fusion in the nucleus, but unlike photons, they rarely interact with matter. Most of the neutrinos produced end up escaping the Sun immediately. For several years, measurements of the number of neutrinos produced by the Sun were three times lower than predicted. This problem was recently solved with the discovery of neutrino oscillation effects.

ALPHA CENTAUR

The Alpha Centauri star is a triple star system located about 4.37 light-years from Earth in the constellation Centaurus. It is the closest star to our solar system, and can be seen with the naked eye in the southern hemisphere.

The system consists of three stars: Alpha Centauri A, Alpha Centauri B, and Proxima Centauri. Alpha Centauri A and B orbit each other, forming a binary system, while Proxima Centauri is farther away and orbits the central pair.
Alpha Centauri A is the brightest star in the system, with a mass slightly greater than that of the Sun, while Alpha Centauri B is slightly smaller and cooler. Proxima Centauri is a red dwarf star, approximately one eighth the mass of the Sun.

There is a lot of interest in Alpha Centauri as a potential destination for space exploration and the search for extraterrestrial life, since it is the closest star to our solar system. Several missions and initiatives are being planned to study this star system more closely.

Each of these stars has its own distinct physical and chemical characteristics.

Alpha Centauri A is a yellow-white star, with a mass about 1.1 times that of the Sun, a radius about 1.22 times the radius of the Sun, and a temperature about 5800 Kelvin. Its luminosity is about 1.5 times that of the Sun.

Alpha Centauri B is a yellow and orange star, with a mass about 0.9 times that of the Sun, a radius about 0.86 times the radius of the Sun, and a temperature of about 5,300 Kelvin. Its luminosity is about 0.5 times that of the Sun.

Proxima Centauri is a red dwarf star, with a mass of about 0.12

times that of the Sun, a radius of about 0.14 times the radius of the Sun, and a temperature of about 3000 Kelvin. Its luminosity is about 0.0015 times that of the Sun.

As for the chemical composition, the three stars are composed mainly of hydrogen and helium, with traces of other elements such as carbon, oxygen, nitrogen, iron and other metals. Analysis of the light emitted by stars allows scientists to determine the chemical composition and other physical properties of these celestial objects.

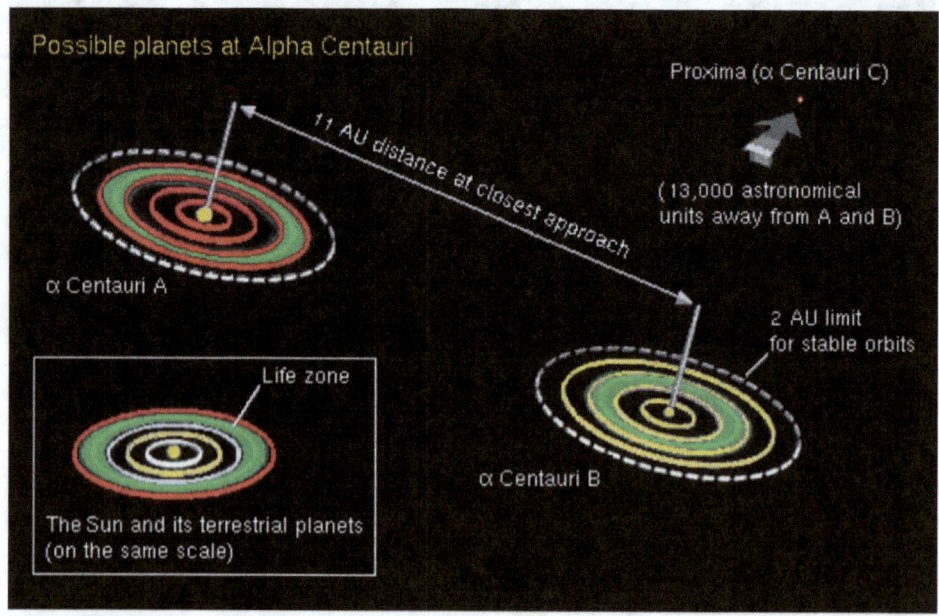

The distance between Alpha Centauri A and Alpha Centauri B varies with time, due to their elliptical orbit around their common center of mass. This distance ranges from about 11 astronomical units (AU) at the periastrum (the closest point in the orbit) to about 35 AU at the apoastrum (the farthest point in the orbit). On average, the distance between the two stars is about 23.7 AU.

The distance between Alpha Centauri A and Proxima Centauri is about 13,000 AU, or about 4.24 light-years. The distance between Alpha Centauri B and Proxima Centauri is about 12,900 AU, or about 4.22 light-years.

In summary, the stars of the Alpha Centauri system are relatively close to each other compared to other stars in the universe, but they are still too far away to reach with current technologies.

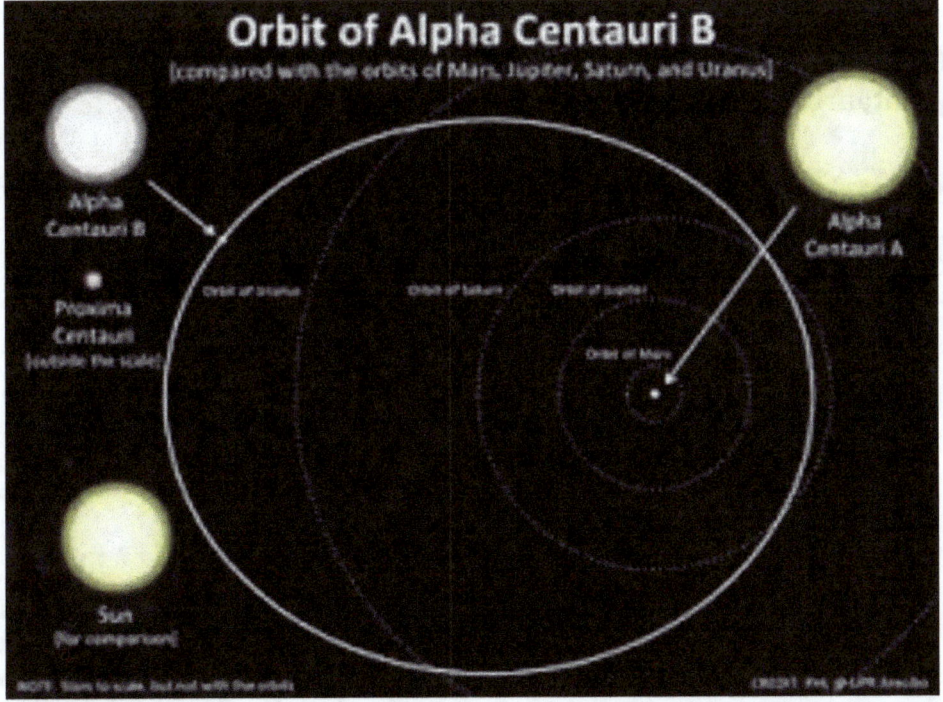

So far, a few planets orbiting stars in the Alpha Centauri system have been discovered, but none of them directly orbit Alpha Centauri A or B stars, which form a binary system.

The first planet discovered in the Alpha Centauri system was Proxima b, in 2016, which orbits the star Proxima Centauri in a very close orbit, with an orbital period of about 11.2 days. Proxima b is a rocky planet with a mass similar to that of Earth and orbits in a habitable zone, which means that there could be liquid water on its surface. However, it remains to be seen if the planet has an atmosphere suitable for supporting life.

In 2017, another planet orbiting the star Alpha Centauri B was discovered, but its existence has yet to be confirmed by other observatories and more research is needed to confirm its presence.

In addition to these two planets, there are several initiatives underway to search for more planets in the Alpha Centauri system, including the "Breakthrough Starshot" project, which proposes sending a fleet of ultra-fast space probes to study the system up close. With these efforts, more planets in the Alpha Centauri system may be discovered in the future.

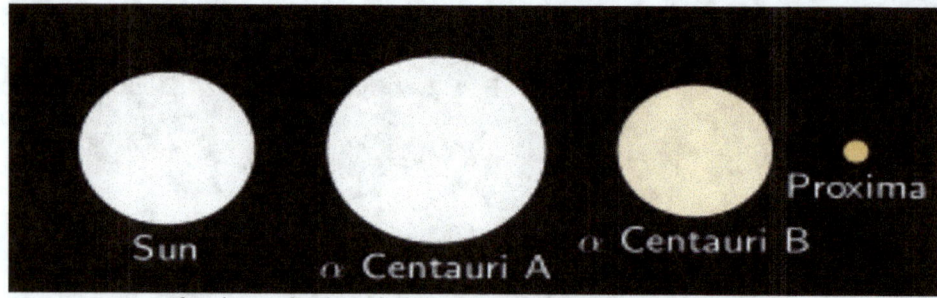

The size and color of Alpha Centauri's components appear to be to scale compared to the Sun.

SIRIUS

Sirius is a binary star located in the constellation Canis Major. It is the brightest star in the night sky, with an apparent magnitude of -1.46. The main star, known as Sirius A, is a main sequence star of spectral type A1V, while the companion, known as Sirius B, is an extremely dense white dwarf. The distance of Sirius from Earth is about 8.6 light years, which makes it one of the closest stars to us, in terms of kilometers, this distance is equivalent to about 8.1 trillion km (8.1×10^{12} km).

That distance is relatively close in astronomical terms, making Sirius one of the closest stars to our solar system. The proximity of Sirius has allowed astronomers to study and observe the star with detail and precision, using different observation techniques such as spectroscopy, photometry and interferometry.

In addition, Sirius is of great historical and cultural importance to many world societies, including the ancient Egyptian culture and the indigenous Dogon culture, who have legends and myths about the star.

The chemical and physical composition of Sirius A, the primary star of the binary system, is well known to astronomers and scientists. Based on spectroscopic observations, the chemical composition of Sirius A is thought to be similar to that of the Sun, composed primarily of hydrogen (about 71% by mass) and helium (about 27% by mass), with trace amounts of other heavy, such as oxygen, carbon, iron, nitrogen and others.

In terms of physics, Sirius A is an A1V star, with an estimated surface temperature of around 9,940 Kelvin and a mass of about 2.02 solar masses. Its luminosity is about 25 times greater than that of the Sun and its age is estimated at about 230 million years. It is a very stable star and is in the main phase of its stellar evolution, converting hydrogen into helium in its core through nuclear fusion reactions.

Sirius B, the companion star of the binary system, is an extremely dense and hot white dwarf, with a mass of about 0.6 solar masses and an estimated radius of only 0.0085 times the radius of the Sun. The temperature of its surface is around 25,200 Kelvin, making it one of the hottest stars known. Sirius B is believed to be the exposed core of a giant star that lost its outer atmosphere earlier in its evolution. The orbital distance between the two stars is about 20 astronomical units (AU).

Composed of two stars that orbit around a common center of mass, due to the gravitational force acting between them, the main star, Sirius A, has a greater mass than the companion star, Sirius B, and therefore the center of mass of the binary The system is closest to Sirius A.

Sirius B's orbit around Sirius A is very small compared to Earth's orbit around the Sun. According to observations, the mean distance between the two stars is about 20 astronomical units (AU) and the orbital period is about 50.1 years. The eccentricity of the orbit is very low, which means that the distance between the stars does not vary much during the orbit.

The gravitational interaction between the two stars has observable effects, such as a periodic change in Sirius A's apparent position in the sky, known as proper motion. In addition, Sirius B's orbit is tilted relative to Earth's line of sight, causing periodic variations in the brightness of the binary, known as radial velocity variations. These variations make it possible to determine the mass and other properties of the stars in the binary system.

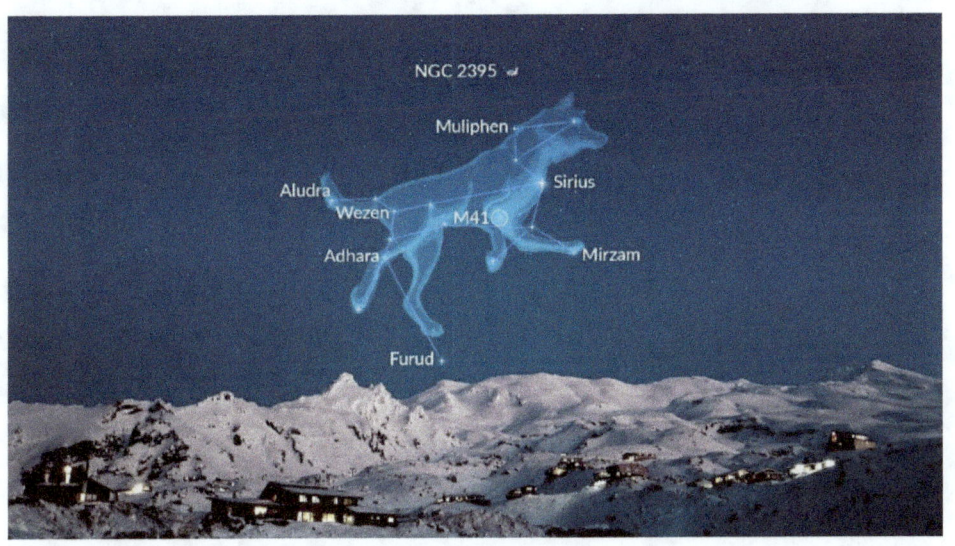

WR 104

The WR 104 star is a binary star system located in the constellation of Sagittarius, about 8,000 light years from Earth. It is classified as a Wolf-Rayet star, a type of extremely luminous and massive star that is nearing the end of its life.

The binary system consists of two stars that orbit around a common center of mass. One of the stars is a Wolf-Rayet star with a mass about 25 times that of the Sun, while the other is a smaller but more massive star with a mass about 10 times that of the Sun.

One of the most interesting features of WR 104 is the presence of a cloud of dust surrounding the stars, which is believed to have been ejected from the system earlier in its evolution. This cloud of dust is believed to be spiral or top-shaped, and could be the precursor to a future supernova explosion.

Due to its location in the Milky Way, WR 104 is heavily obscured by interstellar dust, making it difficult to study. However, we continue to observe the system using a variety of techniques, including infrared and X-ray observations, to learn more about the properties and evolution of massive stars.

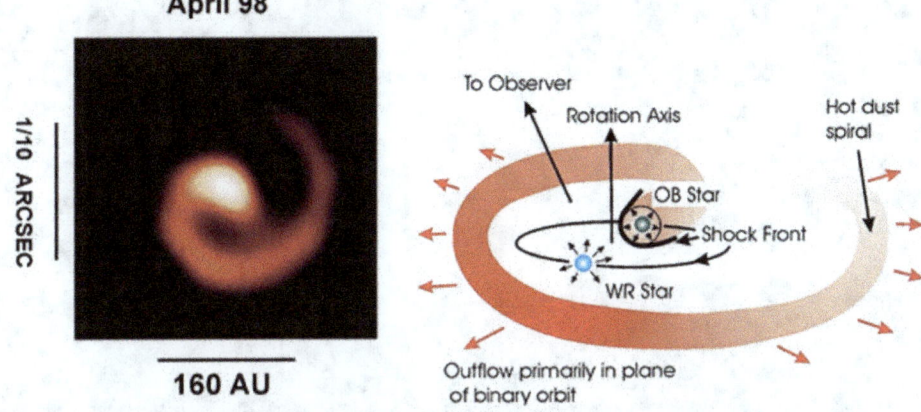

WR 104 at 2.27 Microns
April 98

Interacting Binary Wind Model
of Spiral Outflow Around WR 104

1/10 ARCSEC

160 AU

To Observer
Rotation Axis
Hot dust spiral
OB Star
Shock Front
WR Star
Outflow primarily in plane
of binary orbit

There is no scientific evidence that WR 104 poses a direct risk to Earth. Although it is a massive and unstable star, and could eventually explode in a supernova, the effects of the explosion are unlikely to reach Earth directly due to its distance.

However, a nearby supernova explosion can have collateral effects on Earth, such as increasing cosmic radiation, causing changes in the climate, and affecting the ozone layer. Also, if the dust cloud around WR 104 were to point towards Earth, it could affect the atmosphere and possibly cause a meteor shower.

However, it is important to note that the chance of a supernova occurring at WR 104 is considered very low, and even if it does, the chance of it significantly affecting Earth is greatly reduced.

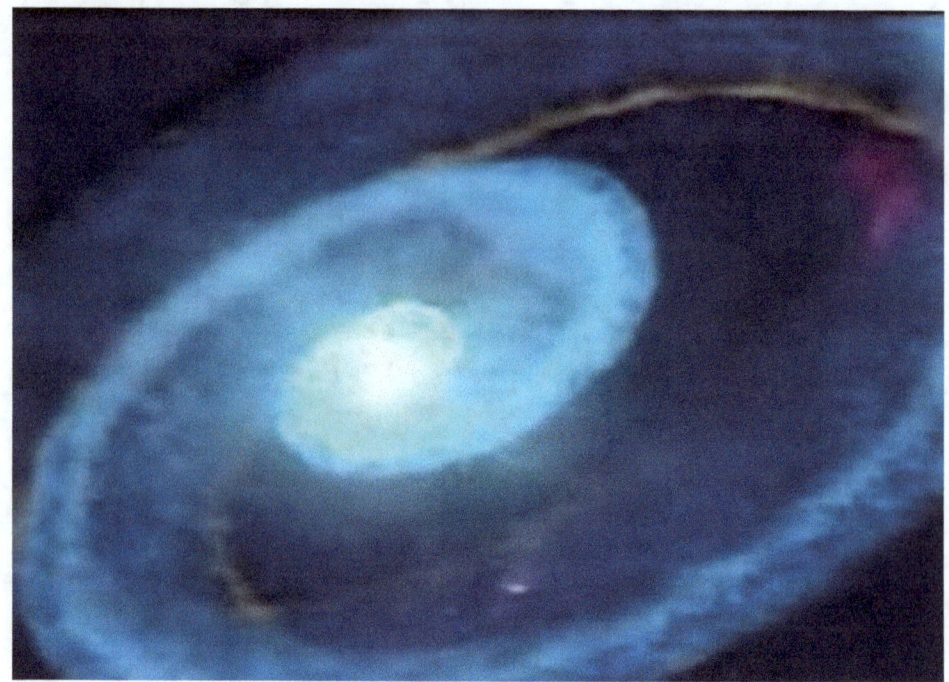

As an extremely massive and hot star, with an estimated surface temperature of 50,000 to 60,000 degrees Celsius, it has shed most of its outer layer of hydrogen and helium through the strong stellar wind, exposing inner layers of higher elements. heavy.
Spectroscopic studies indicate that WR 104 is rich in heavy elements such as carbon, oxygen, nitrogen, silicon, and iron. In addition, analysis of the light emitted by the star suggests the presence of other elements, such as neon, magnesium, sulfur, and argon.

The star is also known to be surrounded by a cloud of dust, probably containing organic and mineral compounds produced by the heavy elements emitted by the star.

Its spectrum shows the presence of a variety of elements, and the surrounding dust cloud contains organic and mineral compounds.

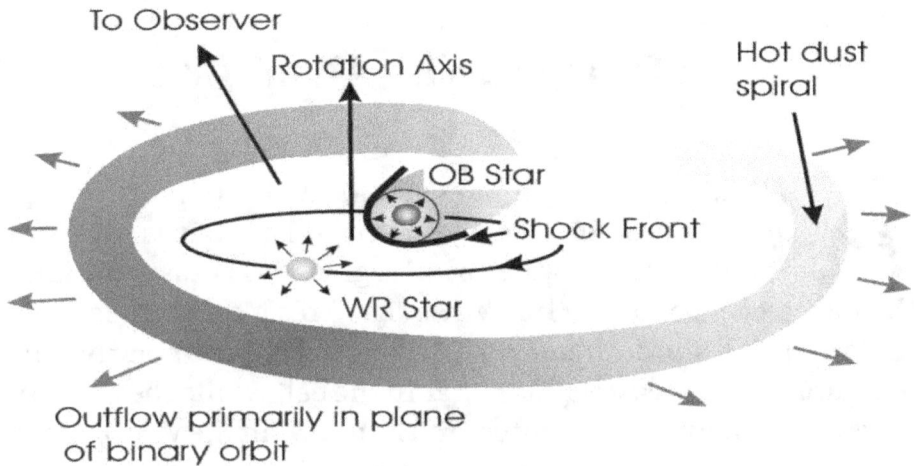

The orbit of the star WR 104 is complex, since the two stars are very close to each other and influence each other with their gravity. The smaller, more massive star orbits the Wolf-Rayet star every 220 days, while the distance between the two stars varies between 10 and 30 times the average distance between Earth and the Sun.

In addition, the inclination of the orbit with respect to the line of sight of the Earth is high, which makes us see the system from an inclined angle, making it difficult to observe and correctly analyze the orbit.

ZETA ORIONIS-ALNITAK

Alnitak is a blue supergiant star located in the constellation Orion, about 800 light years from Earth. It is one of the brightest stars in the Orion region and is easily visible to the naked eye, popularly known as "Las Tres Marías". It is part of "Orion's Belt", a prominent formation of three stars in the night sky. Alnitak is the easternmost star in the belt, while the other two stars are Alnilam (in the center) and Mintaka (in the west). Alnitak has an estimated mass of about 30 times that of the Sun and is a very young star, estimated to be about 6 million years old.

Alnitak has an estimated mass of about 30 times the mass of the Sun and an estimated diameter of about 20 times the diameter of the Sun. This means that Alnitak is an extremely large and bright blue supergiant star with a physical size of about 40 million km. (approximately 28 times the distance between the Earth and the Sun) and a surface temperature of around 28,000 degrees Celsius.

Alnilam is a blue supergiant star located in the constellation Orion, just like Alnitak and Mintaka. It has an estimated mass of about 30 times the mass of the Sun and an estimated diameter of about 36 times the diameter of the Sun. This means that Alnilam is an extremely large star, with a physical size of about 23 million kilometers (about 16 times the distance between them and about 31,000 degrees Celsius). Mintaka is the westernmost star in Orion's Belt, while Alnilam is the belt's central star, and Alnitak is the easternmost star.

Alnitak, Alnilam, and Mintaka are all blue supergiant or blue-white giant stars, which means they have similar chemical and physical compositions. The chemical composition of these stars is primarily determined by nuclear fusion taking place in their cores, which converts hydrogen to helium and produces a variety of heavier elements through further fusion reactions.

From spectroscopic studies, we know that these stars contain hydrogen, helium, and a host of heavier elements, including carbon, nitrogen, oxygen, neon, magnesium, silicon, and iron. In addition, these stars also contain smaller amounts of other elements, such as sodium, aluminum, calcium, and nickel.

Regarding their physical structure, these stars have dense and hot nuclei where the nuclear fusion reactions that generate the energy they radiate take place. These cores are surrounded by layers of ionized gas that form the atmosphere of the stars. The temperature and pressure in these layers decreases as we move further from the core, which leads to the formation of different zones with different physical and chemical properties.

Additionally, these stars also have powerful magnetic fields that can affect their atmospheres and produce phenomena such as stellar winds, solar flares, and other magnetic activity. In short, the stars Alnitak, Alnilam, and Mintaka are complex and fascinating celestial objects that continue to challenge our scientific understanding in many ways.

Stars as massive as these are much shorter-lived than smaller stars like the Sun. They burn up their nuclear fuel at a much faster rate, which means they have a much shorter lifespan.

The stars Alnitak, Alnilam and Mintaka are estimated to be between 5 and 10 million years old. That may sound like a lot, but compared to the age of the universe, which is estimated to be around 13.8 billion years, they are relatively young. These stars are estimated to have a few hundred thousand to a few million years before exhausting their nuclear fuel and collapsing to become neutron stars or black holes.

Orion constellation, image that represents the origin, symbology and mythology.

These three stars do not orbit each other, but orbit the center of the Milky Way along with our Sun and billions of other stars. The orbit of these stars around the center of the Milky Way is mainly influenced by the galaxy's gravity and the distribution of matter in its region.

The orbital velocity of stars in Orion's Belt can be measured from their radial velocity, which is the speed at which they are moving toward or away from us along the line of sight. From these measurements, we estimate that the stars Alnitak, Alnilam and Mintaka move at a speed of about 20 to 30 kilometers per second around the center of the Milky Way, this means that they take about 200 million years to complete one orbit around of the Milky Way. galaxy.

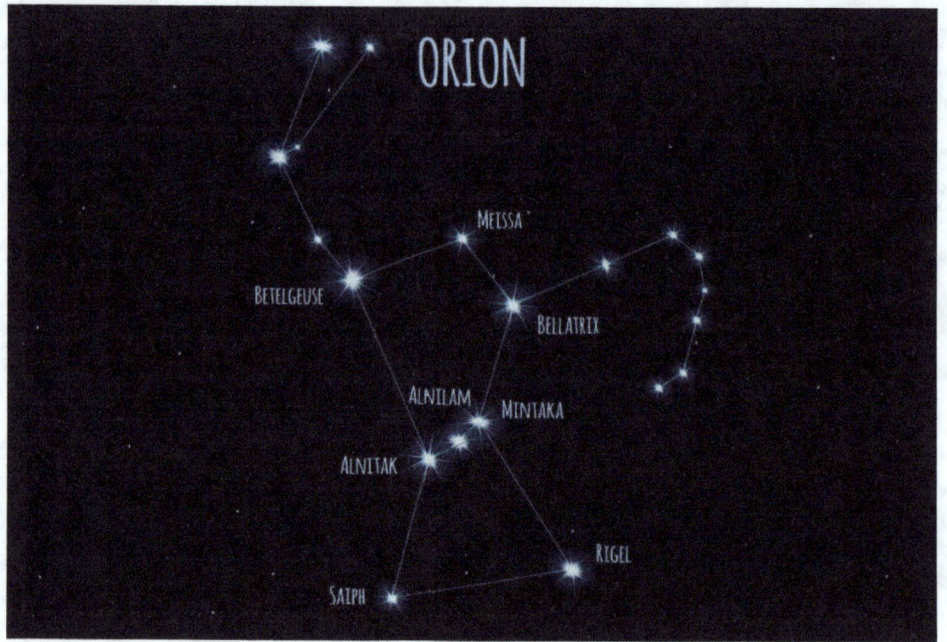

ALDEBARAN

Aldebaran is a red giant star in the constellation Taurus. It is the brightest star in the constellation and the 13th brightest star in the night sky, easily recognizable by its reddish color and its prominent position near the Pleiades star cluster.

The star has an apparent magnitude of 0.85 and an absolute magnitude of -0.63, which means it is about 425 times brighter than the Sun. It lies about 65 light-years from Earth and has an estimated mass of about 1.7 solar masses.

Aldebaran has been important to various cultures throughout history, including the ancient Persians, who believed that the star was the pupil of the heavenly eye. The Arabs called her "the follower" because she seemed to follow the Pleiades across the night sky.

The star orbits around the center of the Milky Way, just like the Sun and other nearby stars. However, as is common in astronomy, Aldebaran's orbit is more easily described in terms of its relationship to the solar system, since this is what we observe from Earth.

Aldebaran is not part of the solar system, but is about 65 light years from Earth. It moves through space with an average speed of about 50 km/s relative to the Sun. Its orbit around the Milky Way is much wider and slower, taking about 625 million years to complete a single revolution around the Sun. galactic center. It is known to have a close binary partner, although this one is much fainter and harder to observe. The companion star orbits

Aldebaran with a period of about 600 years and is at an average distance of about 1,500 million kilometers from the main star.

Its effective temperature is around 3,900 degrees Celsius, much colder than the temperature of the Sun, which is around 5,500

degrees Celsius. As a result, Aldebaran emits most of its light in the infrared range.

Chemically, it consists mainly of hydrogen and helium, like most stars. However, it also contains significant amounts of other elements such as carbon, oxygen, and nitrogen, these elements are created within the star through nuclear reactions that take place in its core and outer layers.

As Aldebaran ages, it undergoes a series of transformations to its internal structure, depleting hydrogen in its core and beginning to burn helium, expanding and cooling in a process known as a red giant. As the helium runs out, the star will continue to evolve and expand further, eventually shedding its outer layers and forming a planetary nebula.

Some fun facts about this celestial body is that in modern Western popular culture, Aldebaran is often quoted in songs, movies, and books as a poetic reference to the night sky and the cosmic nature of the universe. In the science fiction series "Star Trek", Aldebaran is mentioned several times as an important place in the galaxy. For example, the crew of the USS Enterprise visits the planet Aldebaran III in an episode of the original series, and was eventually considered in Persian mythology to be the "ward of the heavenly eye" and one of the four royal stars associated with the four items. of the nature. Aldebaran represented the element of fire.

CRUCIS RANGE

The star Gamma Crucis, also known as Gacrux, is one of the brightest stars in the constellation of the Southern Cross, located in the southern celestial hemisphere. It is one of the four stars that make up the famous Southern Cross asterism, which is one of the most iconic symbols of the southern night sky.

Gacrux is an M-class red giant star with a surface temperature of around 3,500 Kelvin. It is an LC-type variable star, which means that its luminosity varies slightly over time. Its apparent magnitude varies between 1.59 and 1.66, making it easily visible to the naked eye even in urban areas with polluted skies.

With an estimated mass of about 1.5 times the mass of the Sun and a diameter of about 120 times the diameter of the Sun, Gacrux is a very large star. Its luminosity is about 1,500 times that of the Sun, making it one of the brightest stars in the Universe.

Gacrux is relatively young, with an estimated age of around 25 million years. Although it is relatively close to Earth in astronomical terms, at a distance of about 88 light-years, not much is known about its planetary systems or exoplanets. However, the discovery of planets around other M-class stars suggests that Gacrux may have at least one planetary system orbiting it.

Gacrux is an important star to the indigenous people of Australia, who know him as "Gnokan Danna" or "Heaven's Gate Guardian". It is one of the most sacred stars in the Australian night sky and plays an important role in many Aboriginal stories and myths.

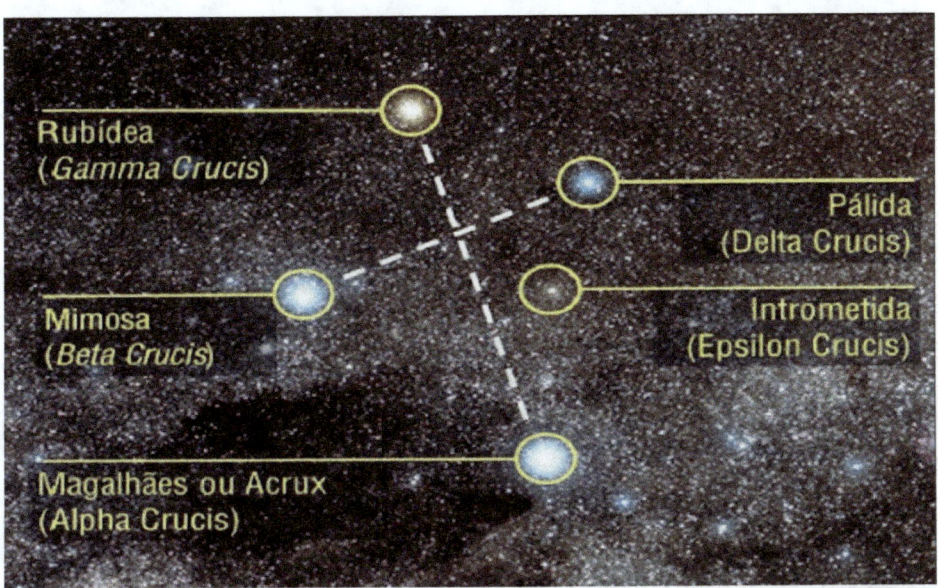

Rubídea
(*Gamma Crucis*)

Pálida
(Delta Crucis)

Mimosa
(*Beta Crucis*)

Intrometida
(Epsilon Crucis)

Magalhães ou Acrux
(Alpha Crucis)

Regarding its internal structure, Gacrux has a nucleus that is surrounded by a shell of ionized hydrogen, followed by a shell of ionized helium, and finally a shell of neutral hydrogen. The outer shell of the star is composed mainly of gas and dust, which are expelled from its surface during stellar evolution.

Gacrux is a low-mass star, which means that its internal structure is different from that of more massive stars. The energy is generated mainly by the fusion of hydrogen into helium in the core of the star, and convection is responsible for transporting this energy to the surface. Convection is a process in which hot gas rises to the star's surface, while cooler gas falls toward the core.

In summary, Gacrux is an M-class star with a simple chemical composition, composed mainly of hydrogen and helium. Its internal structure is different from that of more massive stars, with energy generated mainly by the fusion of hydrogen into helium in the core and transported to the surface by convection.

Gacrux orbits around the center of the Milky Way, the spiral galaxy in which our solar system is located. Its orbit is determined by the gravity exerted by other objects in the galaxy, including stars, clouds of gas and dust, and dark matter.

According to astronomical observations, Gacrux has a radial velocity relative to the Sun of about -19.7 km/s, which means that it is moving away from us at that speed. Its space velocity is estimated to be about 22 km/s, indicating that it moves in an eccentric orbit around the center of the Milky Way.

Gacrux's position in the sky gradually changes over time, due to its motion around the center of the galaxy. The entire path of the star around the center of the Milky Way takes about 250 million years to complete, known as its orbital period.

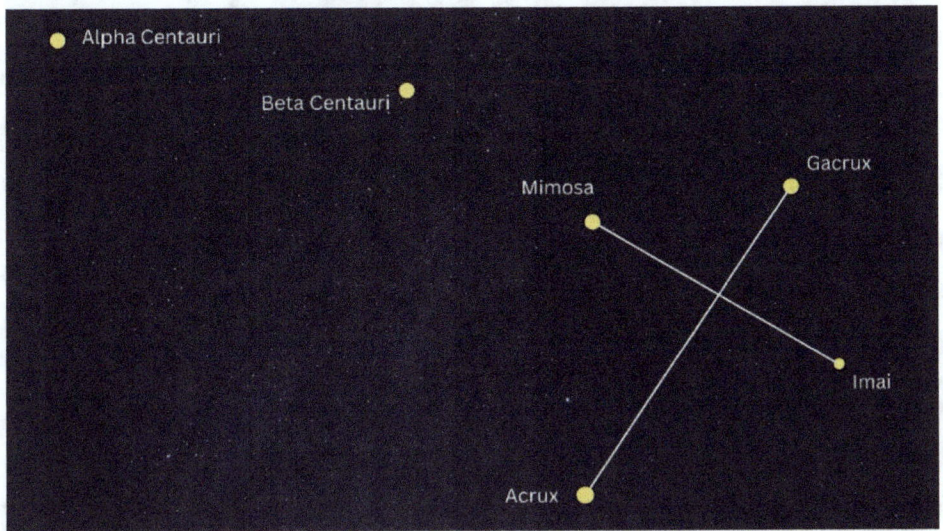

Due to its relative proximity, Gacrux is often used as a reference for measuring distances to other stars and celestial objects in the galaxy.

A curious fact is the study of this star and other nearby ones, which are important for understanding the formation, evolution and composition of the stars in our galaxy.

ETA CARINAE

Eta Carinae is a star located in the constellation of Carina or (Quilla), about 7,500 light years from Earth. It is one of the brightest stars in the night sky and has been the subject of intense study by astronomers over the years.

The star Eta Carinae is classified as a luminous blue variable star and was discovered in 1677 by astronomer Edmond Halley. Since then its luminosity has fluctuated and in 1843 it experienced one of the largest stellar explosions ever recorded, temporarily becoming the second brightest star in the night sky.

The 1843 stellar explosion released an enormous amount of energy and created two huge clouds of gas, called Homunculus and Weigelt Haze, which expanded at speeds of up to 1,500 km/s. The Homunculus is an hourglass-shaped bipolar nebula that surrounds the star, while the Weigelt Haze is a series of concentric rings that surround it.

Since the explosion, Eta Carinae has decreased in brightness and size, but remains a massive and unstable star. It is estimated that it has a mass of about 100 times that of the Sun and a luminosity of more than five million times that of the Sun. Its surface temperature is around 25,000 degrees Celsius.

Eta Carinae is believed to be nearing the end of its lifespan and could soon explode in a supernova. Although the star is a safe distance from Earth, an explosion of this magnitude could affect Earth's atmosphere and cause significant damage to communication systems.

Eta Carinae continues to be an important source of study with advanced observation techniques such as space telescopes and interferometry to study its structure and behavior. We need more data to understand this star, which continues to challenge scientists' understanding of the nature of the universe.

Image credits: NASA

The chemical composition of this star is complex and is not yet fully understood by scientists. However, spectroscopic studies suggest that Eta Carinae is a star rich in heavy elements such as carbon, nitrogen, oxygen, and iron, indicating that it has already gone through several stages of nuclear fusion at its core.

Furthermore, the star is known to have a high proportion of helium in its atmosphere, suggesting that it is a young star that has not yet had time to convert all the helium into heavier elements through nuclear fusion processes. This high proportion of helium could also be a sign that Eta Carinae is a star that formed from primordial gas with low metal content.

Other elements detected in Eta Carinae's atmosphere include silicon, magnesium, sulfur, and argon. However, the relative abundance of these elements is not yet fully known.

Image credits: NASA

Eta Carinae does not have an orbit in the traditional sense of the word, since it is a single star and not in a binary or multiple system. However, the star is known to exhibit variations in its luminosity and other properties, which can be explained by cycles of stellar activity, including oscillations in its internal structure and periodic flares.

Furthermore, the star lies at the inner edge of a large star-forming region called the Carina Nebula, which contains several young and massive stars. The gravitational interaction between these stars may play an important role in the evolution of Eta Carinae and its stellar activity.

Although it does not have a defined orbit, Eta Carinae's position in the sky is known with precision and is often used as a reference point for astronomical navigation. The star is located in the constellation Carina and can be seen with the naked eye under good viewing conditions.

However, more recent studies state thatbe abinary star systemvery close to each other. the lesser stardiameteris the hottest (30,000 °C) and the other with three times thediameterit is colder (15,000 °C) but twice as bright. Thisstar systemis wrapped in a densecloudingasesIsdust, which forms a nebula 400 times larger than theSolar system, known as theEta Carinae Nebula(or NGC3372). The loss of luminosity is possibly due to a consequence of the closer approach between the two stars, theperiastrom, at which point the smaller star covers almost half of the larger one. The decrease in brightness is equivalent to 20 times that ofSun, but shining like 4 to 5 million suns. The period of rotation of the stars (with respect to each other) is 5.5 years.

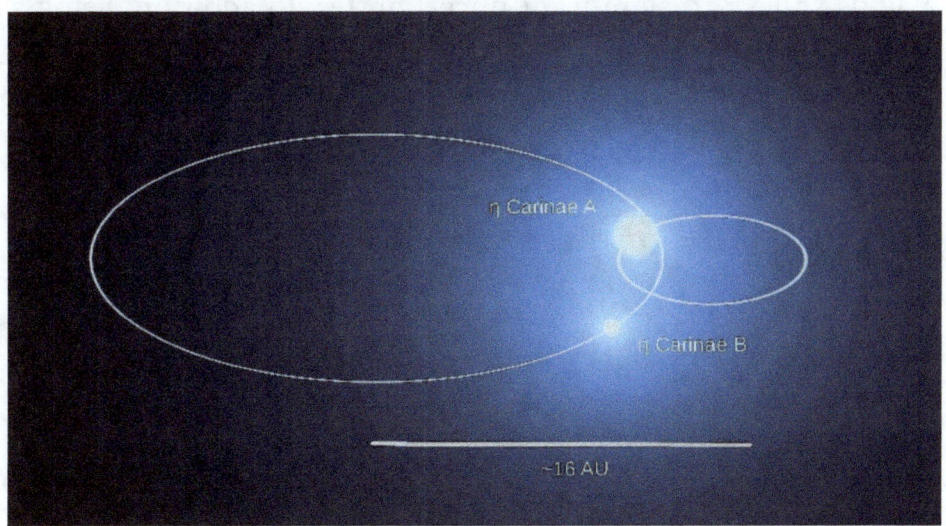

The Brazilian astronomer Augusto Damineli, a professor at IAG-USP, is one of those who affirm that the star is a variable because every five and a half years, according to him, there is a reduction in its brightness, since other astronomers did not accept it. this theory, in the However, in 1997, there was a further reduction in brightness, the phenomenon was confirmed. In 2003, thanks to the records of more than 50 specialists backed by observations through terrestrial and orbiting telescopes, it was finally confirmed that it was indeed another variable star of the SDOR type - Binary High Luminosity Stars, with variations between 1 to 7 magnitudes, associated with and enveloped in expanding material typical of nebulae.

Very large stars like Eta Carinae run out of fuel very quickly due to their disproportionately high luminosity. Eta Carinae is expected to explode as a supernova or hypernova in the next few million years.

And finally, andstudies suggest that Eta Carinae rotates very slowly, with an estimated rotation period of about 5.5 years. However, this estimate is based on indirect measurements and may be subject to significant uncertainties. In addition, being a variable and unstable star, it is difficult to calculate its rotation with precision.

BETELGEUSE–APHA ORIONIS

It is one of the most famous and easily recognizable stars in the night sky. Located in the constellation Orion, it is the second brightest star in that constellation, second only to Rigel. However, it is one of the brightest stars in the night sky and is about 100,000 times more luminous than the Sun.

One of Betelgeuse's most notable features is its size. It is estimated to have a diameter about 1,000 times that of the Sun, making it one of the largest known stars. If it were placed at the center of our solar system, its atmosphere would extend beyond the orbit of Jupiter.

Another characteristic that makes it interesting is that it is a variable star, which means that its luminosity changes over time, due to its magnitude, these changes can be easily detected with the naked eye. On average, it takes about 420 days for the star to complete a full cycle of brightness. The variation in brightness is caused by the pulsation of the star, which causes changes in its temperature and luminosity.

It has recently drawn media attention due to speculation about its possible explosion in a supernova. Betelgeuse is at the end of its life and is expected to eventually explode in a supernova. However, there is no certainty when this will occur. Some studies have suggested that the star could explode at any moment, while others claim that it still has thousands of years before it explodes.

Regardless of when the star explodes, its death will be a significant event for astronomy. The explosion will be visible from Earth and can be seen even during the day, depending on how light is scattered through the atmosphere. In addition, the supernova will produce an incredible amount of energy and matter, which can be studied by astronomers for many years.

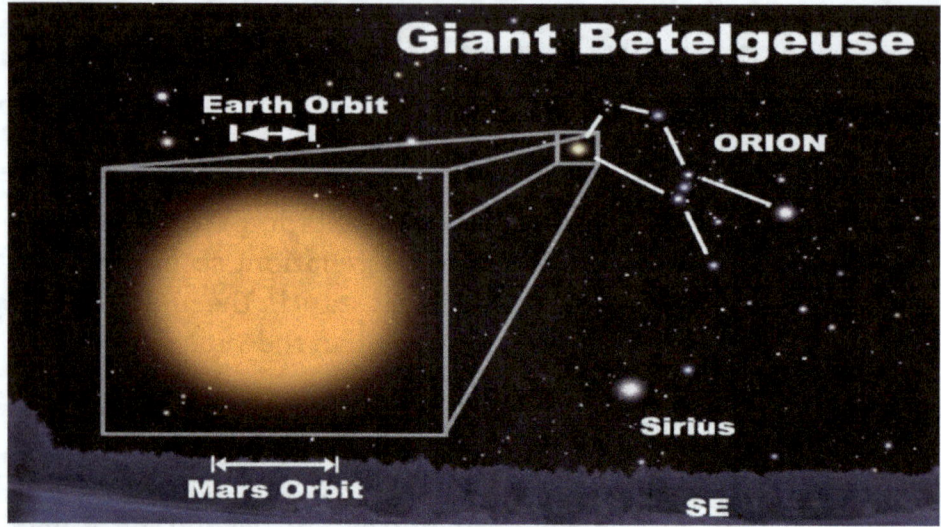

Betelgeuse is a very large, luminous, and cool star classified as a red supergiant of spectral type M1-2 Ia-ab. The letter "M" indicates that it is a red star belonging to the spectral class M, which is why it has a low surface temperature; the suffix "Ia-ab" is the luminosity class of the star and indicates that it is intermediate between a normal luminosity supergiant and a high luminosity supergiant. The main feature of the visual spectrum of stars of this type is the presence of titanium(II) oxide (TiO) absorption bands in the green region of the spectrum, indicating low surface temperature. The low intensity of the neutral calcium line at 4227 Å is the main indicator of high luminosity. Since the introduction of the MKK rating system in 1943,

Red supergiants like Betelgeuse are massive stars that have already exited the main sequence and are in the late stages of their evolution. These stars burn up their fuel quickly and live for only a few million years. Originally a main sequence O-class star, Betelgeuse has already consumed all the hydrogen in its core, causing the core to contract under the force of gravity. To balance out the hotter, denser core, the outer layers expanded and cooled.

While its exact evolutionary status is unknown, Betelgeuse is most likely fusing helium to generate carbon and oxygen in the core, with a shell of hydrogen fusion surrounding the core.

Artist's representation of the star and itsnebula

The most abundant elements in Betelgeuse's atmosphere are hydrogen and helium, which make up approximately 85% and 13% of the chemical composition, respectively. The other elements present are mainly carbon, oxygen, nitrogen, silicon, sulfur, iron and titanium, among others.

The star is thought to have evolved from a very massive

star, which produced many heavier elements through nuclear reactions in its core. These heavier elements were then transported to the star's surface via convective processes in its atmosphere.

As far as the orbit is concerned, Betelgeuse does not orbit any specific object. Instead, it is a lone star moving through the Milky Way along with other stars. It moves in a relatively random trajectory, affected mainly by gravitational interactions with other stars and massive objects in the galaxy.

In terms of rotation, Betelgeuse has a relatively slow rotation, with a rotation period of about 8.4 years. That's surprisingly slow for a star of its mass and size, estimated to be about 20 times the mass of the Sun and about 1,000 times the size of the Sun. Betelgeuse's slow rotation is thought to be due to interactions between rotation and the outer layers of the star, which are highly convective.

ANTARES

Antares is a red supergiant star located in the constellation of Scorpio. With an estimated diameter of about 700 times that of the Sun, Antares is one of the largest known stars. Its distance from Earth is approximately 550 light years, making it one of the brightest stars in the night sky.

The name "Antares" comes from the Greek ant-Ares, which means "the rival of Mars." This is because the star has a reddish hue similar to that of the red planet.

Antares is a very hot star, with a surface temperature of around 3,500 degrees Celsius, but its red color is the result of its large size and the emission of light at longer wavelengths.

In addition to its impressive appearance, Antares is also quite a complex star. It is known to have a binary star system, which means there is another star orbiting close to it, Antares's companion star is much smaller and cooler than it is, and it takes about 900 years to complete one orbit around the star major.

It is an evolved star, with an estimated age of about 12 million years, it has already gone through the phase in which it produces energy through the nuclear fusion of hydrogen into helium, and now it is in the phase in which it is converting the helium into carbon and oxygen into its core. This evolution will eventually lead to the death of the star, but since Antares is so much larger than the Sun, its death will be much more dramatic.

At the end of its life, Antares will explode in a supernova, an extremely powerful explosion that will release an enormous amount of energy and matter into space. This can create a phenomenon known as a planetary nebula, which is a cloud of gas and dust illuminated by radiation from the dying star. Despite not being close enough to pose a direct threat to Earth, the Antares explosion would certainly be an impressive sight for astronomical observers.

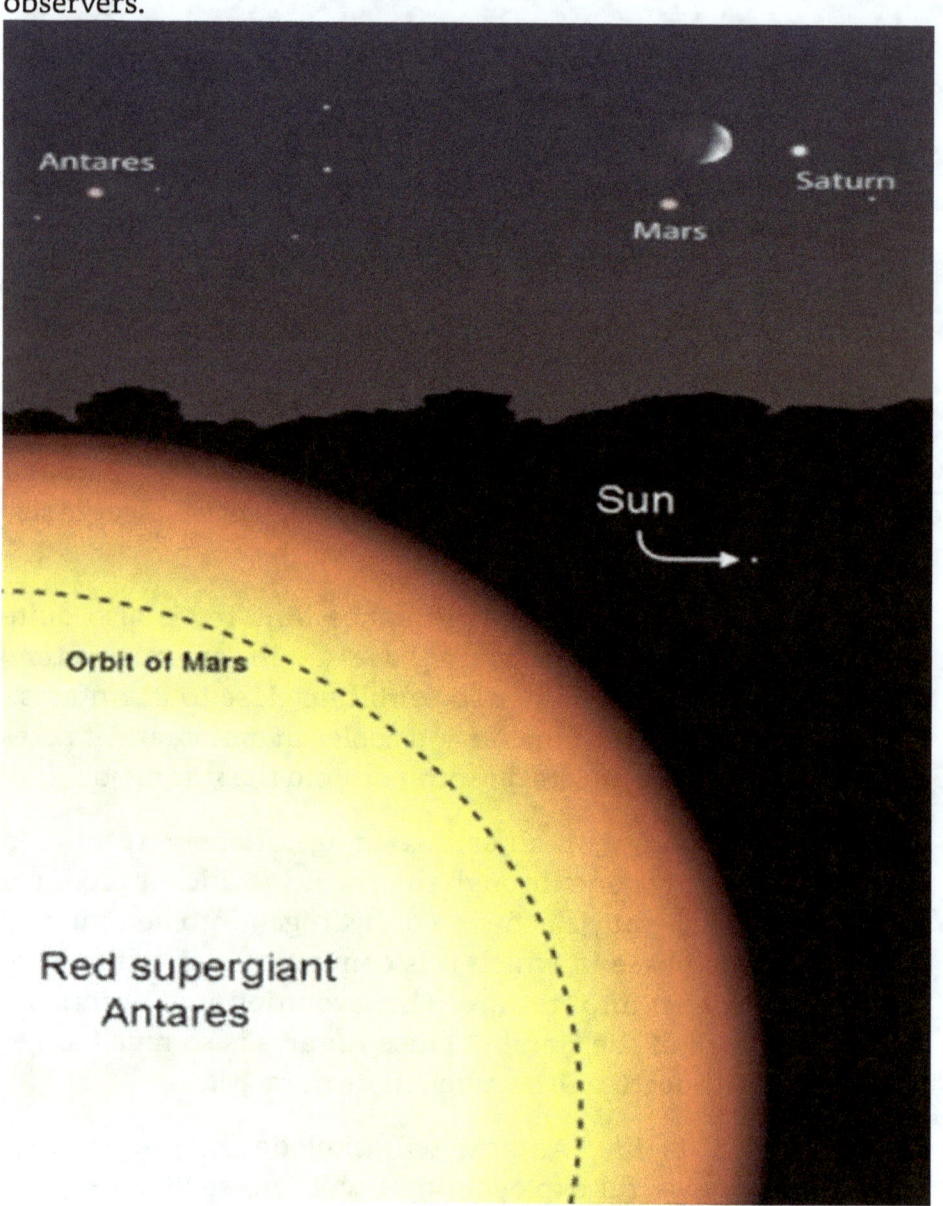

The chemical composition of Antares is quite similar to that of other supergiant stars, it is composed mainly of hydrogen and helium, with trace amounts of heavier elements.

The star produces energy through nuclear fusion, which occurs in the core of the star. During nuclear fusion, the nuclei of atoms fuse to form new nuclei, releasing a large amount of energy in the process. Nuclear fusion of hydrogen into helium is the main source of energy for stars, including Antares.

In addition to hydrogen and helium, Antares contains trace amounts of other chemical elements such as carbon, oxygen, nitrogen, and iron. These elements are formed in nuclear reactions that occur within the star as it evolves.

The amount of heavier elements in Antares is relatively small compared to the amount of hydrogen and helium. That's because supergiant stars like Antares are very young in cosmic terms and haven't yet had enough time to produce large amounts of the heavier elements through nuclear reactions.
However, even small amounts of the heavier elements in stars like Antares are important for planet formation and life itself. Most of

the chemical elements found on Earth, including carbon, oxygen, and iron, were formed in stars that existed before our Sun. When these stars exploded in supernovae, they released these elements into space, which were subsequently clump together to form new stars and planets.

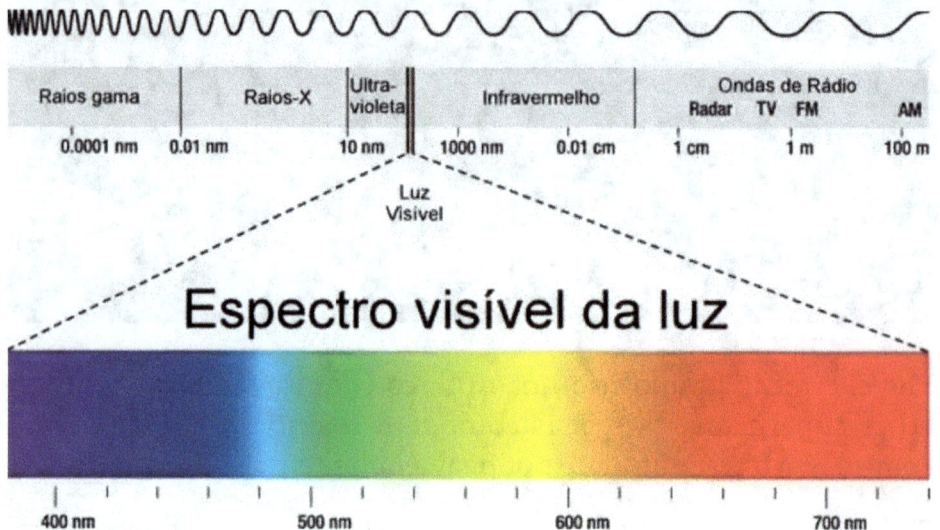

MU CEFEI

The star Mu Cephei, also known as the red giant star or simply "Mu Cep", is one of the brightest known stars in the Milky Way. Located in the constellation Cepheus, about 2,300 light-years from Earth, it is one of the most massive and luminous stars known, with an apparent magnitude of about 4.08.

Mu Cephei is a class M star, which means that it is a red giant star with a relatively low surface temperature and very high luminosity. It is also a semi-irregular variable, which means that its luminosity varies with time, albeit unpredictably. Its magnitude varies between 3.4 and 5.1, with an average period of about 730 days.

The star Mu Cephei has an estimated mass of about 20 times that of the Sun and a radius of about 1,500 times that of the Sun, making it one of the largest known stars. Its surface temperature is relatively low, around 3,500 degrees Celsius, which makes it red in color. The star has a luminosity about 300,000 times that of the Sun, making it one of the brightest stars known.

Mu Cephei is a very young star, with an estimated age of around 10 million years, which is very young compared to the Sun, which is around 4.6 billion years old. The star has a large amount of circumstellar material, indicating that it is in an active evolutionary phase. The star is thought to eventually become a planetary nebula star, shedding its outer layers in a cloud of gas and dust.

Its great mass and luminosity make it an important example for understanding stellar evolution in extremely massive stars.

Furthermore, the star is an important source of infrared radiation and is used to study dust formation around red giant stars.

The chemical composition of the star Mu Cephei is well studied by astronomers and astrophysicists around the world, and is known to be very different from the chemical composition of the Sun.

Spectroscopic analyzes indicate that the star has a very low abundance of elements heavier than helium, known as "metals" in astronomy. The ratio of iron to hydrogen, for example, is only about 0.06% of the solar ratio. This suggests that the star Mu Cephei is a second population star, which formed from very old, metal-poor gas.

This star has an excess of carbon over oxygen, suggesting that the star underwent deep convective mixing at some point in its evolution. This process may have occurred when the star fused helium into carbon and oxygen in its core and then transported these elements to the star's surface layers.

Other chemical elements detected in the star include hydrogen, helium, lithium, carbon, oxygen, nitrogen, sodium, magnesium, aluminum, silicon, sulfur, calcium, titanium, and iron. The chemical composition of the star Mu Cephei is important for understanding stellar evolution in second-population stars and for comparing it with the chemical composition of other stars in the Milky Way.

The orbit of the star Mu Cephei is not well known, as it is a solitary star and has no known stellar companion. However, studies can estimate the star's radial velocity, which is the speed at which it is moving away from or toward Earth, based on the Doppler shift of the spectral lines in its spectrum. This can provide information about the star's average orbital velocity relative to the center of the Milky Way.

The radial velocity of the star Mu Cephei is relatively low, about 14.5 km/s relative to the Sun. This suggests that the star is orbiting the center of the Milky Way in a relatively circular orbit, since stars with orbits more ellipticals generally have more variable radial velocities.

As for the rotation of the star Mu Cephei, astronomers believe that the star probably has a very slow rotation, since red giant stars usually have very slow rotations due to the expansion of their outer layers. The star's rotation can be estimated from the width of the spectral lines in its spectrum, which are wider in the most rapidly rotating stars. However, these spectral lines in red giant stars are often very wide due to the star's low surface temperature, making it difficult to accurately measure the star's rotation.

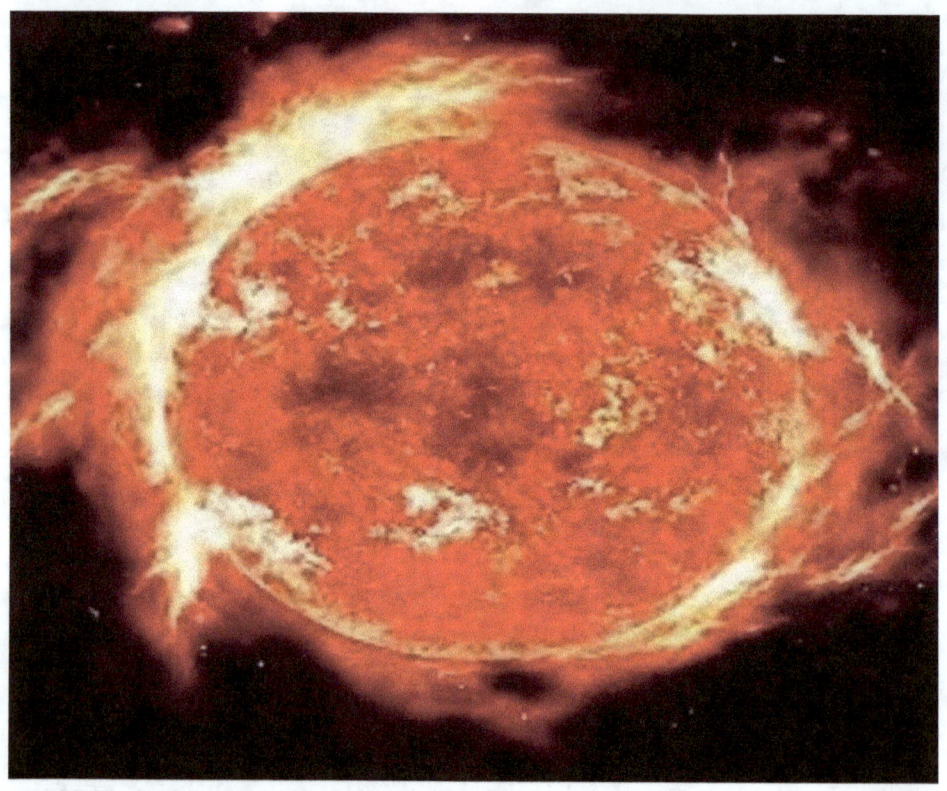

VY CANIS MAJORIS

T he star VY Canis Majoris is one of the most fascinating and enigmatic stars ever discovered. Located in the constellation of Canis Major, about 1.2 KPC (Kiloparsecs) from Earth, this star is one of the largest and most luminous known to man. In this chapter, we will explore the characteristics, history of discovery, and mysteries surrounding VY Canis Majoris.

Discovery and characteristics of VY Canis Majoris;

VY Canis Majoris was discovered in 1801 by Jérôme Lalande, a French astronomer, while conducting a survey of stars. At that time, Lalande listed the star as the twenty-second brightest in the constellation Canis Major.

Today we know that VY Canis Majoris is a supergiant red variable star that is entering an advanced phase of its stellar evolution. It is classified as a star of spectral type M and has an estimated mass of about 20 times that of the Sun.

The diameter of VY Canis Majoris is enormous, some 2,000 times that of the Sun. If it were at the center of our solar system, its radius would extend as far as the orbit of Jupiter. Its volume is equal to about 5 billion times the volume of the Sun. To get an idea of the magnitude of this star, if VY Canis Majoris were placed in our solar system, the distance between it and the Earth would be only half that the distance between the Sun and Pluto.

VY Canis Majoris is also one of the most luminous stars in the known universe, emitting luminous energy some 500,000 times that of the Sun. However, this enormous luminosity is emitted mainly in the infrared, meaning the star is dimmer. in the visible spectrum.

Mysteries and curiosities about VY Canis Majoris

VY Canis Majoris is such a large and complex star that scientists still do not fully understand how it works. One of the big questions is how such a big star manages to stay stable, since the star's gravitational pull would have to be so strong that it would collapse in on itself. Furthermore, the star is emitting an enormous amount of material, including dust and gas, raising questions about how this is possible in such a massive star.

Another curiosity about VY Canis Majoris is that it is a variable star, which means that its luminosity changes over time, on some occasions the star has become brighter than any other known star, while on others it has dimmed making it almost invisible. . .

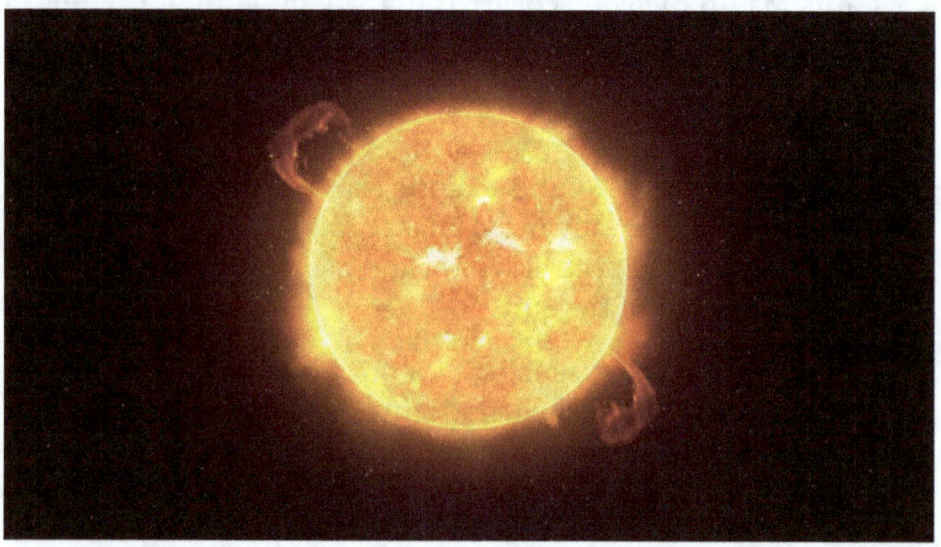

Another interesting curiosity about VY Canis Majoris is that it emits a large amount of material, between dust and gas, that spreads through the space that surrounds it. Astronomers believe that this material is the result of intense stellar activity on the star's surface and that it is undergoing a phase of intense mass loss.

The orbit of VY Canis Majoris is somewhat difficult to define, since the star is solitary and has no close stellar companion. However, scientists were able to determine that it is moving towards the center of the Milky Way, our galaxy, at a speed of about 22 km/s. In addition, it is considered a high-velocity star, which means that it moves relative to our Solar System at a speed much greater than the average for the stars in the galaxy.

Regarding the rotation of VY Canis Majoris, it is important to note that red supergiant stars rotate very slowly relative to smaller, younger stars. This is because these stars have a highly expanded atmosphere, which means that the star's surface is very far from the core, where the rotation takes place. Furthermore, the rotation of such a massive star would be very difficult to measure precisely using current observing techniques.

However, some studies have indicated that it may be slowly spinning around its axis. A 2015 study, for example, suggested that the star could be rotating at a speed of only 1 km/s, which is extremely slow compared to the speed of the Sun's rotation, which is around 2 km/s.

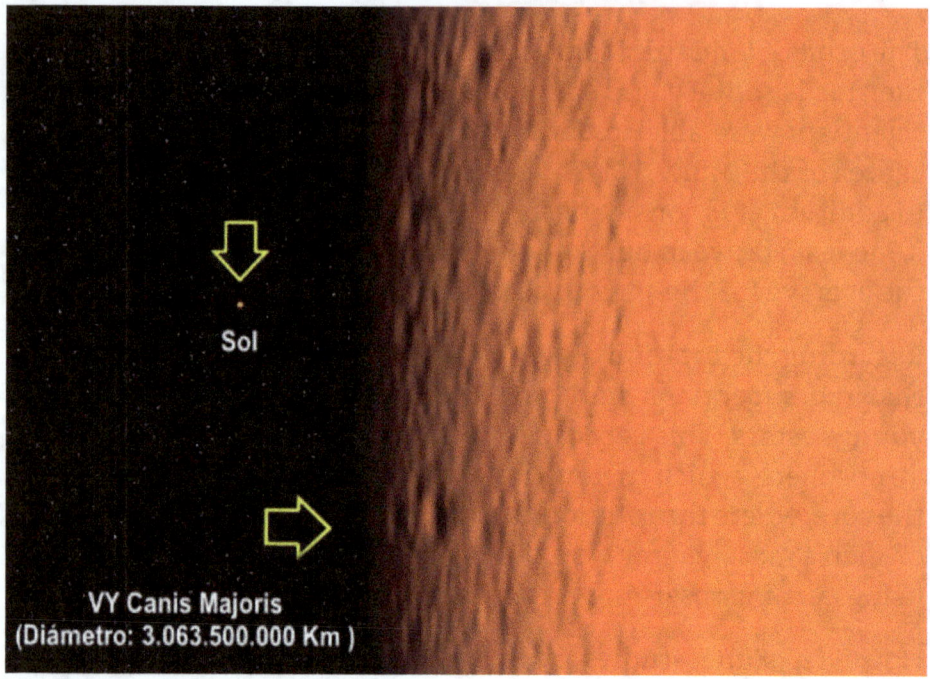

Sol

VY Canis Majoris
(Diámetro: 3.063.500.000 Km)

The chemical composition of VY Canis Majoris is similar to that of other red supergiant stars, with a mixture of light elements such as hydrogen and helium and heavier elements such as carbon, oxygen and iron. However, due to its size, the star also contains elements that are relatively rare in other stars, such as technetium and lithium.

Furthermore, VY Canis Majoris is known to be a variable star, which means that its luminosity and surface temperature fluctuate over time. This can affect the chemical composition of the star, since the nuclear reactions that take place in its core can be different at different times. In fact, some studies suggest that VY Canis Majoris may be undergoing a process of heavier element fusion at its core, which could lead to significant production of even heavier elements.

Regarding the physics of VY Canis Majoris, it is a very large star, with an estimated radius of around 1,800 times the radius of the Sun, due to this magnitude the star has a very low surface gravity, which allows its atmosphere to it expands. far beyond the core of the star. This expanded atmosphere is responsible for many of the observed characteristics of the star, such as its low surface temperature and high level of luminosity.

R. W. CEFEI

The star RW Cephei, also known as V712 Cephei, is a variable star located in the constellation Cepheus. It is one of the most luminous stars known in the Milky Way, with an apparent magnitude ranging from 5.7 to 11.5. The star is classified as a red supergiant and belongs to the spectral class M3-M5.

The first mention of RW Cephei was made in 1895 by the American astronomer Edward Pickering, who included it in a list of variable stars. Since then, the star has been widely studied and monitored by astrophysicists and astronomers from around the world.

The main feature that makes RW Cephei so interesting is its variability. Its apparent magnitude varies irregularly in periods that can last from a few days to a few decades. Short-term cycles of variation (lasting from a few days to a few weeks) are caused by pulses of expansion and contraction of the star, while long-term cycles (lasting decades) can be caused by changes in the internal structure of the star or by the influence of a companion star.
In addition to variability, other interesting features of RW Cephei include its mass, radius, and temperature. Recent estimates suggest that the star's mass is about 25 times that of the Sun, while its radius is about 1,200 times that of the Sun. This means that if the star were placed in the place of the Sun, it would extend beyond the orbit of the Sun. Jupiter's temperature is relatively low for such a massive star, with an effective temperature of around 3,500 K.

The star is also known to be a source of radio emission. The radio emissions are caused by electrons being accelerated in magnetic fields in the star's atmosphere. Recent studies suggest that RW Cephei may be generating a source of X-ray emission, possibly due to interaction with a companion star.

In terms of stellar evolution, RW Cephei is nearing the end of his life. Red supergiants are known to experience thermonuclear explosions, which can cause the ejection of their outer atmosphere and the formation of planetary nebulae. However, RW Cephei has yet to show any imminent signs of a thermonuclear explosion.

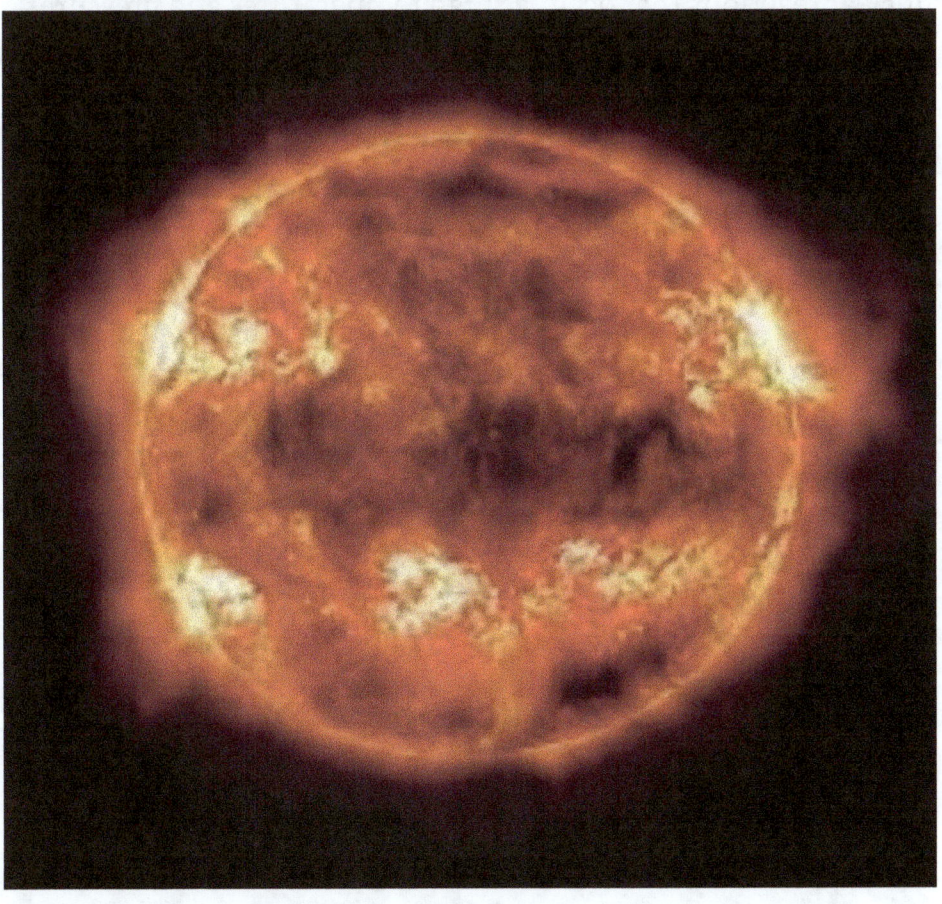

RW Cephei is located at a distance of approximately 4 KPC (Kiloparcescs) from Earth. This distance is very large and makes direct observation of the star difficult, but astronomers can study it with the help of telescopes and sensitive instruments, such as space telescopes. This distance from Earth is one reason why much remains to be discovered about this star and other red supergiants. Astronomy continues to develop new technologies and techniques to overcome distance challenges and learn more about these fascinating and complex stars.

In terms of chemical composition, RW Cephei is a star that is extremely rich in heavy elements such as carbon, oxygen, and metals. These elements are produced inside the star through nuclear reactions that occur at high temperatures and pressures.

It is also known to have a large amount of dust in its atmosphere. This dust is made up of microscopic grains of solid material, such as silicates and graphite, that form in the star's outermost layers. The presence of dust can affect the way the star emits light and can cause variations in its luminosity over time.

In addition, RW Cephei is a star known for its strong stellar winds, these winds are formed by charged particles that are thrown at high speeds from the star's surface. Stellar winds are responsible for transporting material from the star to the interstellar medium, contributing to the formation of new stars and planets.

Because it is a solitary red supergiant star, it means that it does not orbit any other stars. It is located in the Milky Way and moves in a trajectory around the galactic center together with other stars.
The orbital speed of RW Cephei is influenced by the distribution of mass in the galaxy, including the mass of dark matter, which astronomers do not yet know.

Regarding the rotation, it is known that the red supergiants have a low rotation rate, this is because these stars have a very thick and expanded atmosphere, which causes the rotation of the star to slow down due to friction between them. the outer layers of the star and the interstellar medium. . Furthermore, the presence of strong magnetic fields can further affect the rotation of the star.

The rotation of stars is an important parameter for understanding how they evolve over time, and RW Cephei's low rotation rate is an important factor to consider in studies of its evolution and behavior. Precise observations of the star's radial velocity can be used to estimate its rotation rate, but this can be difficult due to the complexities of the star's thick atmosphere and the limitations of currently available observing techniques.

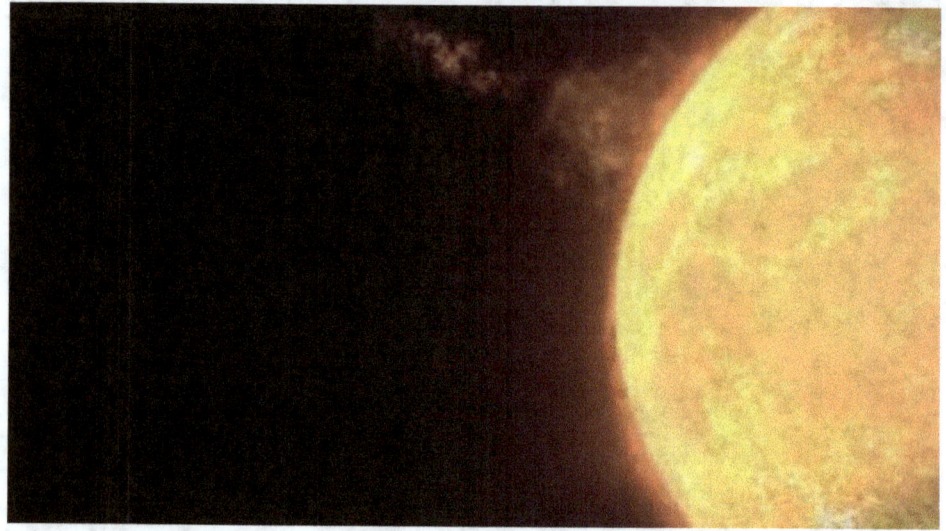

POLAR STAR

The Pole Star, also known as the North Star or Polaris, is a star visible from Earth's northern hemisphere that plays a key role in astronomical navigation and orientation. In this chapter, we will discuss the North Star in detail, including its location, history, physical features, and cultural significance.

The Pole Star is an F7 class star located in the constellation Ursa Minor. It is visible from anywhere north of the equator, and as such is an important reference star for navigators and astronomers alike. The position of the North Star is quite stable, which makes it a reliable tool for determining the direction of north. However, the North Star is not the brightest star in the night sky, but it is relatively easy to identify since it is the closest star to the point where all the lines of longitude meet.

The history of the Polar Star goes back thousands of years. In ancient Greece, the star was known as "Phoenice," meaning "phoenix," and was seen as a symbol of renewal and resurrection. In Norse mythology, the North Star was associated with a goddess named Frigg, who was seen as the guardian of the sky and stars. In Chinese culture, the North Star was known as "Zhen", meaning "true north", and was seen as a symbol of guidance and stability.

North Star's physical characteristics are also quite interesting. It is a yellow-white star, with an apparent magnitude of about +2.0. In terms of size, it is about 6 times larger than the Sun and has a surface temperature of around 6,000 degrees Celsius. Polar Star is also a double star, made up of two smaller stars that orbit around each other.

The Pole Star has been used for astronomical navigation for

centuries. Throughout history, people have used the star to determine the direction of north, aiding land and sea navigation. With the invention of the astrolabe and the sextant, the North Star became even more useful for navigation.

Stars like Polaris form from clouds of interstellar gas and dust that collapse under their own gravity. When the core of this cloud becomes dense and hot enough, it begins to fuse hydrogen into helium, starting the process of nuclear fusion. During this process, energy is released and a series of nuclear reactions take place, creating heavier chemical elements.

The chemical composition of the North Star is determined by the spectral analysis of the light it emits. This technique involves scattering light from the star into a spectrum of colors, which can be used to determine which chemical elements are present in the star, and in what quantity. The chemical elements that make up North Star include hydrogen, helium, carbon, nitrogen, oxygen, neon, magnesium, silicon, sulfur, iron, nickel, and other heavier elements.

Hydrogen is the most abundant element in the North Star, with around 71% of its total mass. Helium is the second most abundant element, with about 27% of its total mass, the other chemical elements are present in much smaller quantities, with less than 1% of its total mass.

The chemical composition of the Pole Star is important because it helps us understand how stars evolve. As a star ages and depletes its nuclear fuel, it begins to fuse heavier elements together, creating new chemical elements in the process.

These elements are then released into space when the star explodes as a supernova, enriching the interstellar medium with new chemical elements. Analyzing the chemical composition of stars like North Star helps us better understand how chemical elements are created and distributed throughout the universe.

According to the most recent measurements, North Star is located about 434 light-years from Earth. This means that the light emitted by the star takes about 434 years to reach us.

The determination of the distance to the Pole Star was carried out by various astronomical techniques. One of the most widely used techniques is stellar parallax.[6]. Using this technique, astronomers were able to measure the distance to North Star with an accuracy of around 1%.

Regarding its orbit, the Pole Star is a solitary star, that is, it has no close companions. It orbits around the center of the Milky Way, along with our Sun and billions of other stars. Its orbit takes about

25.4 million years to complete and its velocity relative to the center of the galaxy is about 19.5 km/s.

Regarding its rotation, it is a slowly rotating star, it rotates around its own axis in about 25.4 days, which is relatively slow compared to other similar stars. This slow rotation can be explained by the advanced age of the star, where it is estimated to be about 70 million years.

It is worth mentioning that the Pole Star has its position very close to the North Celestial Pole, which is the imaginary point in the sky around which the stars seem to revolve due to the rotation of the Earth.

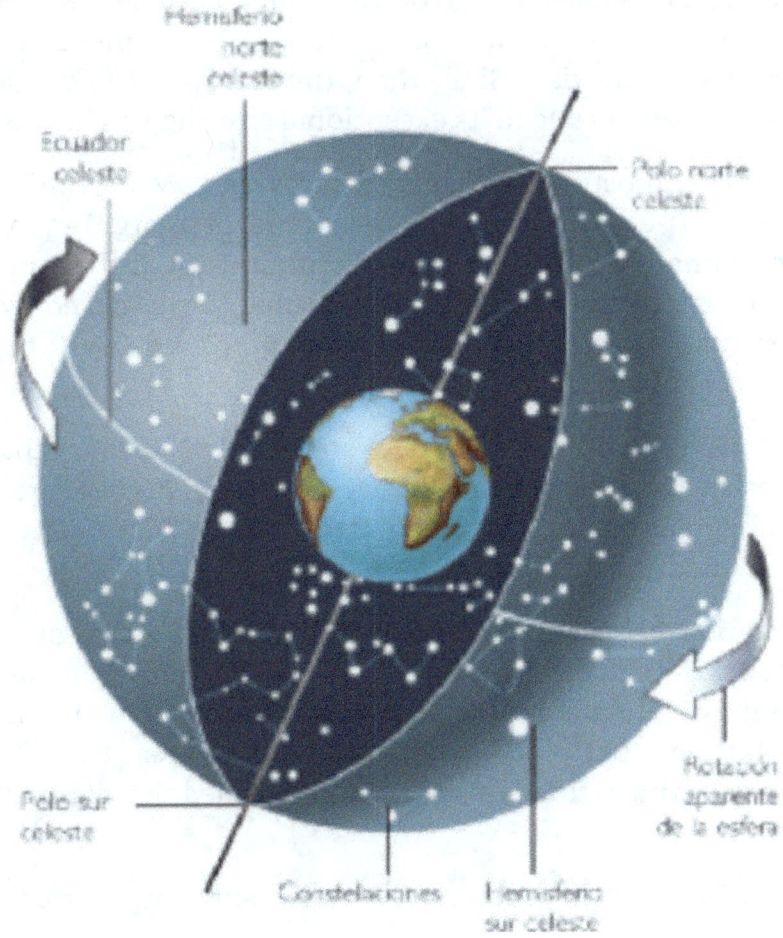

CYGNI NML-V1489 CIGNI

The star NML Cygni is one of the largest and brightest stars known to man. Located in the constellation Cygnus, about 1.6 KLP (kiloparsecs) from Earth, it is a red supergiant star with an estimated radius of about 1,800 times the radius of the Sun.

Discovered in 1965 by a team of astronomers led by Neugebauer, Martz and Leighton, NML Cygni takes its name from the discoverers' last initials. Since then, the star has been studied by many astronomers due to its exceptional size and brightness.

One of the most remarkable features of NML Cygni is its brightness. It emits an enormous amount of energy, equivalent to approximately 500,000 times the luminosity of the Sun. This makes it one of the brightest stars visible to the naked eye. Its temperature is also quite high, reaching around 3,300 degrees Celsius on the surface.

Also, NML Cygni is a variable star, which means that its luminosity and temperature change over time. It goes through a cycle of regular pulses, with a period of about 940 days, which may influence its future evolution.

Astronomers believe that this star is in the final stages of its life, which means that it is running out of fuel at its core. This causes it to lose mass, and it is estimated that it is losing about a millionth of a solar mass per year. This mass loss is so great that the star could be ejecting a cloud of gas around it, called the circumstellar envelope.

Cygni NML could also have important implications for

understanding star formation and stellar evolution. Astronomers are studying the star to try to understand how supergiant stars form and evolve, and how stars like NML Cygni could eventually explode as supernovae.

The chemical composition of the star is not completely known, since it is difficult to obtain precise information about its inner layers. However, from spectroscopic studies, astronomers have some information about the elements present in the star's atmosphere.

NML Cygni is classified as a red supergiant star, which means that it is rich in hydrogen and helium, the most abundant elements in the universe. In addition, other elements such as carbon, oxygen, nitrogen, iron and silicon were detected, although in much smaller quantities.

The heavier elements, such as iron and silicon, are usually produced in the core of stars through nuclear reactions that occur during nuclear fusion.

However, in supergiant stars like NML Cygni, these elements can be produced in the star's outer layers through a process called nucleosynthesis.[7]convective

Also, as it is in the final phase of its life, it may be undergoing chemical enrichment processes, such as convection of heavier material from the inner layers to the outer layers of the star. These processes can lead to a variation in the chemical composition of the star over time.

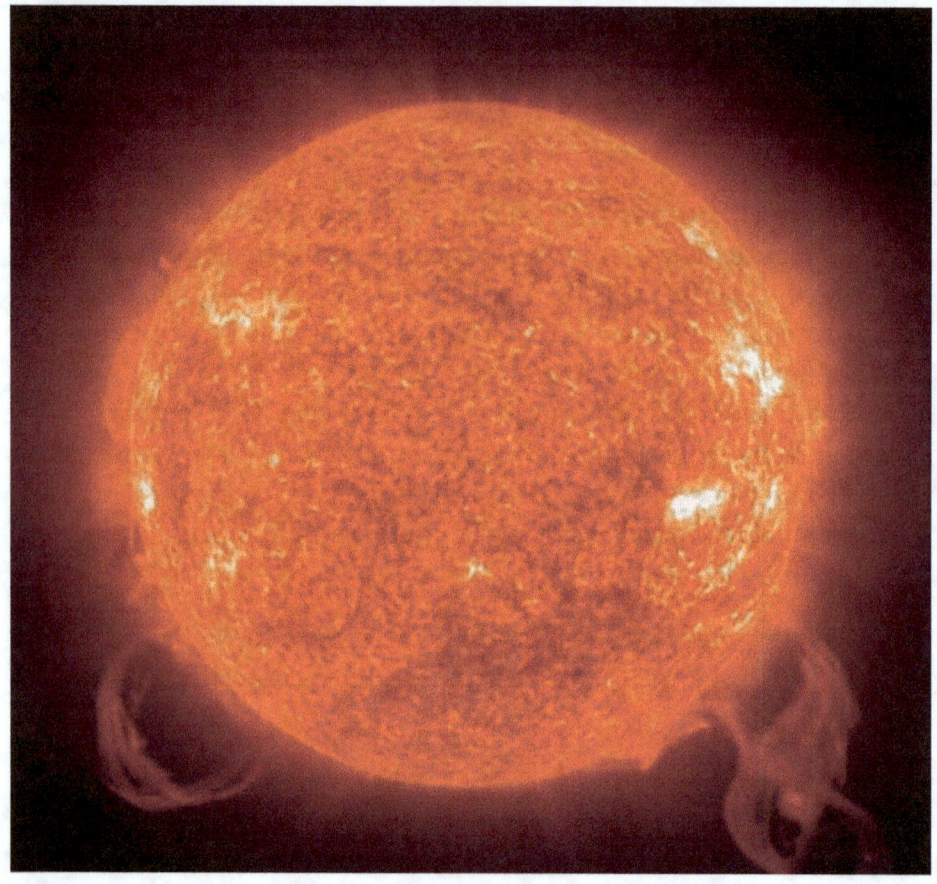

The star's orbit is not precisely known, as it is a great distance from Earth and has no known star system. Therefore, it is difficult to determine its orbit relative to other stars or celestial bodies.

As for rotation, the NML Cygni is known to have a very slow rotation. As a red supergiant star, it has a very large diameter and therefore a longer period of rotation. Estimates indicate that the rotation speed is less than 5 km/s, much slower than the Sun's rotation speed, which is about 2 km/s at the equator.

Importantly, due to its large mass and size, the internal gravitational forces in NML Cygni can also affect its rotation, causing the star to slow down over time.

This information is important for understanding stellar evolution and the behavior of stars at different stages of their lives.

WESTERLUND 1-26

T he Westerlund 1-26 star is one of the most interesting and mysterious stars known to astronomers. Located in the central region of the Carina Nebula, at an approximate distance of 3.52 klp (kiloparsecs) from Earth, this red supergiant star has aroused the curiosity of scientists around the world due to its peculiar characteristics.

Westerlund 1-26 was discovered in 1961 by the Swedish astronomer Bengt Westerlund, who identified it as a very bright and unusual star. Since then, several studies have been carried out to better understand its characteristics and properties.

One of the main features of the Westerlund 1-26 is its size. With an estimated diameter of about 1,500 times that of the Sun, it is one of the largest known stars, classifying it as a red supergiant. Furthermore, it is extremely luminous, with an apparent magnitude of around 12, making it easily visible through powerful telescopes.

Another peculiarity of Westerlund 1-26 is its high temperature. Studies indicate that its surface temperature can reach 20,000 degrees Celsius, making it one of the hottest stars known. This high temperature is associated with its luminosity, since it emits a large amount of energy in the form of visible and ultraviolet radiation.

Furthermore, Westerlund 1-26 is also an unstable star, which means that its luminosity and temperature fluctuate over time. This instability is related to its age, which is relatively young in astronomical terms, around 3 million years. During this time it has gone through several evolutionary phases, such as the fusion of heavier elements in its core and the expansion of its atmosphere.

Another aspect that has drawn the attention of astronomers is the possibility that Westerlund 1-26 harbors a neutron star in its interior. This hypothesis is based on observations indicating that it is surrounded by a ring-shaped nebula, which may have been formed by a supernova explosion. If confirmed, this discovery would be of great importance for understanding the physics of neutron stars and star formation processes in general.

The chemical composition of the star Westerlund 1-26 is a very important aspect to understand its characteristics and evolution. However, the available information on the chemical composition of this star is limited and has not yet been fully determined.

According to some studies, this star is considered to be very metal-rich, which means that it contains a relatively high amount of heavy elements in its atmosphere. Some chemical elements that have been identified in its atmosphere include hydrogen, helium, carbon, nitrogen, oxygen, silicon, and iron.

Spectroscopic observations of Westerlund 1-26 suggest that it has a greater abundance of iron relative to hydrogen than the Sun, which may indicate that it formed from metal-enriched gas. Another piece of information, the presence of carbon in its atmosphere, indicates that it may have gone through a convective mixing process, in which the heavier elements are transported from the core to the surface.

However, current observations do not provide a clear picture of the chemical composition of Westerlund 1-26. Further study is needed to gain a fuller understanding of the abundance of chemical elements in this star and how it may have evolved over time.

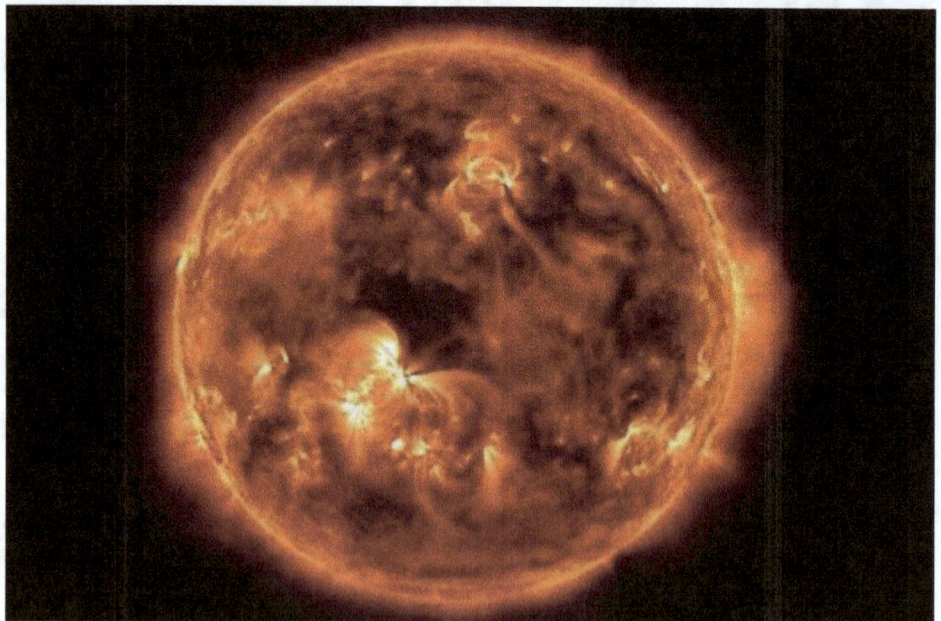

The orbit of the star Westerlund 1-26 around the center of the Carina Nebula has not yet been precisely determined. This is because it is in a very dense and turbulent region, making it difficult to obtain accurate observations. In addition, the star is in a very compact star cluster, which makes determining its orbit even more difficult.

Regarding the rotation, studies indicate that it has a slow rotation, with an estimated equatorial speed of about 20 km/s. This is relatively low for a star with an extremely large size and an estimated mass of around 20 solar masses.

Westerlund 1-26's slow rotation rate can be explained by the fact that it may have undergone tidal coupling with a companion star at some point in its evolution. This process occurs when two stars are close enough that the gravity of one affects the shape of the other, causing their rotations to synchronize.

Another relevant factor is the presence of a strong magnetic field on its surface, which may also be contributing to slow rotation. This is because the star's magnetic field can exert a force that blocks the star's rotation, preventing it from spinning any faster.

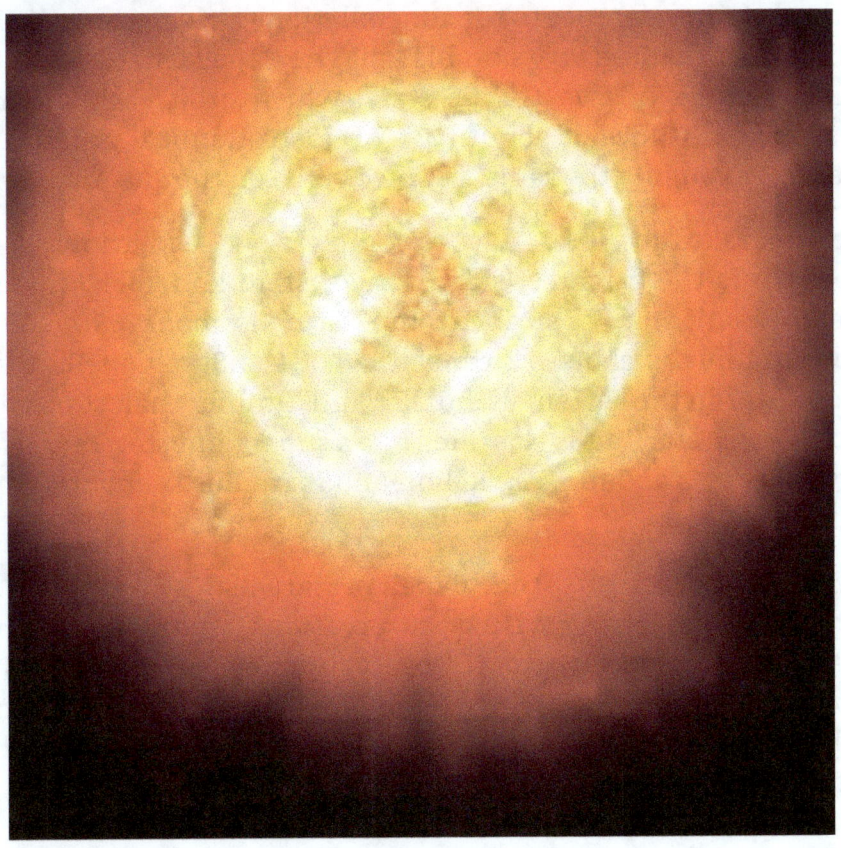

ALPHA AURIGAE (CAPELLA)

The Capella star is a double star located in the constellation Auriga, located about 42 light years from Earth. It is one of the brightest stars in the night sky, with an apparent magnitude of about 0.1. Capella is a yellow giant star that is about 2.5 times more massive than the Sun and about 10 times more luminous. The star is visible to the naked eye and has been one of the most studied stars by astronomers.

The Capella star received its name from a Latin word meaning "little goat," in reference to the constellation Auriga, which depicts a charioteer holding goats on his lap. The Capella star is a double star composed of two G-type stars, which orbit each other at an average distance of about 0.74 AU (astronomical units). This distance is roughly the same distance between the Sun and Venus.

The orbit takes about 104 days to complete one revolution.
Capella A is the brightest star in the system and is classified as a yellow giant star. Its surface temperature is about 4,800 Kelvin and its radius is about 12 times that of the Sun. Capella B, the second star in the system, is smaller and dimmer than star A. It is also a G-type star, but is classified like a supergiant star. Its surface temperature is about 5,500 Kelvin and its radius is about 8 times that of the Sun.

Astronomers studied the Capella star using a variety of techniques, including visual observations, spectroscopy, and interferometry. Spectroscopic observations have shown that the Capella A and B stars are very similar in chemical composition and age, suggesting that they formed and evolved together. Interferometric observations revealed that Capella A has an extended atmosphere, which is expected for a giant star.

The Capella star has been used as a reference point for navigation for centuries. It was one of four stars known as "the nautical stars", which were used to help sailors orient themselves at sea. Additionally, Capella is often used as a calibration star in astronomical studies, due to its known luminosity and relative proximity to Earth.

Spectroscopic and interferometric observations have revealed a wealth of information about the star, including its chemical composition, age, temperature, and size. The Capella star is an important object for both astronomy and navigation, and is an excellent example of how astronomers study and understand stars.

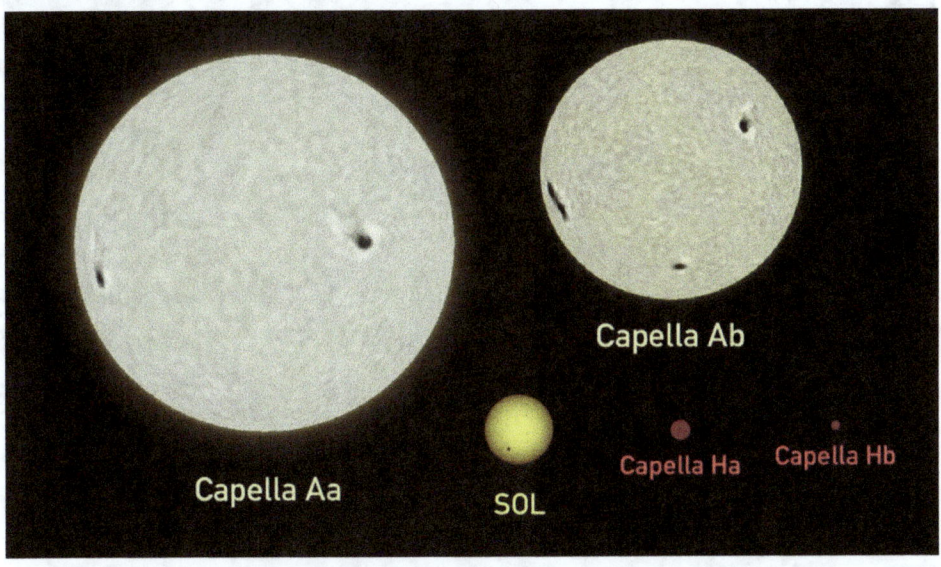

Furthermore, Capella is a very interesting star system to study stellar evolution. Although stars A and B are very similar in chemical composition and age, they have different sizes and temperatures, suggesting that they evolved differently. G-type stars are known to go through a phase where they become red giants, expanding to such an extent that they can swallow nearby planets. Studying Capella could help astronomers better understand how stars evolve and what the consequences of that evolution are.

Spectroscopic studies of the light emitted by stars have revealed that they are composed primarily of hydrogen and helium, the most abundant elements in the universe. In addition, trace amounts of other, heavier elements have been detected in their atmospheres, including carbon, nitrogen, oxygen, iron, silicon, magnesium, and others.

RMC 136A1

T he star RMC 136a1 is one of the most remarkable stars in our galaxy, the Milky Way. Located in the Tarantula Nebula in the Large Magellanic Cloud, RMC 136a1 is one of the most massive and brightest stars known, with an estimated mass of approximately 315 times the mass of the Sun. In this chapter we will present the main features of the star RMC 136a1, as well as its role in stellar evolution.

Its physical characteristics show that it is a Wolf-Rayet star, a class of very massive and hot stars that have lost much of their outer layers of hydrogen. The effective temperature of the star is estimated to be around 50,000 Kelvin, making it one of the hottest stars known. Furthermore, the star has an extremely high luminosity, around 8.7 million times the luminosity of the Sun.

RMC 136a1 is a binary star, which means it is made up of two stars orbiting each other. The companion star is estimated to be about 25 times the mass of the Sun and orbits the parent star in a period of about 20 days.

This star plays an important role in stellar evolution, especially in the formation of black holes. As a very massive star, RMC 136a1 evolves rapidly and exhausts its nuclear fuel on a relatively short time scale compared to less massive stars. When that happens, the star collapses and explodes as a supernova, leaving behind a stellar remnant.

In this case, the supernova explosion will likely result in the formation of a black hole. In addition, RMC 136a1 is also a major source of ionizing radiation in the Tarantula Nebula, making it important for understanding the formation and evolution of HII regions, which are regions of ionized hydrogen.

The chemical composition of the star RMC 136a1 is a constantly evolving area of research and is not yet fully understood. However, studies indicate that the star has a chemical composition that is relatively rich in heavy elements such as carbon, oxygen, nitrogen, silicon, and iron.

THE SUN

Through analysis of the star's spectrum, the astronomers were able to determine that RMC 136a1 has a relatively low helium abundance compared to less massive stars. Furthermore, the star also has a relatively high abundance of nitrogen, which is consistent with its classification as a Wolf-Rayet star.

Spectral analysis also suggests that the star RMC 136a1 may be enriched in heavy elements produced in supernovae, which is consistent with its large mass and rapid evolution. However, further study is needed to fully understand the star's chemical composition and how it relates to its stellar evolution.

UY SCUTI

The UY Scuti star is a fascinating astronomical object that has aroused great interest among the scientific community and the general public. It is a red supergiant located in the constellation of Scutum, whose physical characteristics place it among the largest known stars in the universe.

According to current estimates, UY Scuti has a mass about 30 times that of the Sun and a radius about 1,700 times that of the Sun. These measurements, however, are still subject to some uncertainty, due to the difficulty in obtaining precise observations of stars so far away. The distance in relation to Earth is approximately 2912.65 parsecs, which means that the light emitted by this star takes more than 9 thousand years to reach us.

Spectral analysis of UY Scuti has revealed the presence of various chemical elements in its atmosphere, in addition to hydrogen and helium, such as carbon, oxygen, iron, and other heavy metals. These elements are produced through nuclear reactions in the core of the star and are transported to the surface by convective processes.

Little is known about UY Scuti's orbit around the center of the Milky Way, but it is believed to move in an elliptical orbit, taking millions of years to complete one complete revolution. Regarding the rotation of the star, the observations indicate that it is a low-speed star, which takes about 740 days to complete a complete rotation around its axis. This value is quite unusual for a star of this size, and the causes of this phenomenon are not yet fully understood.

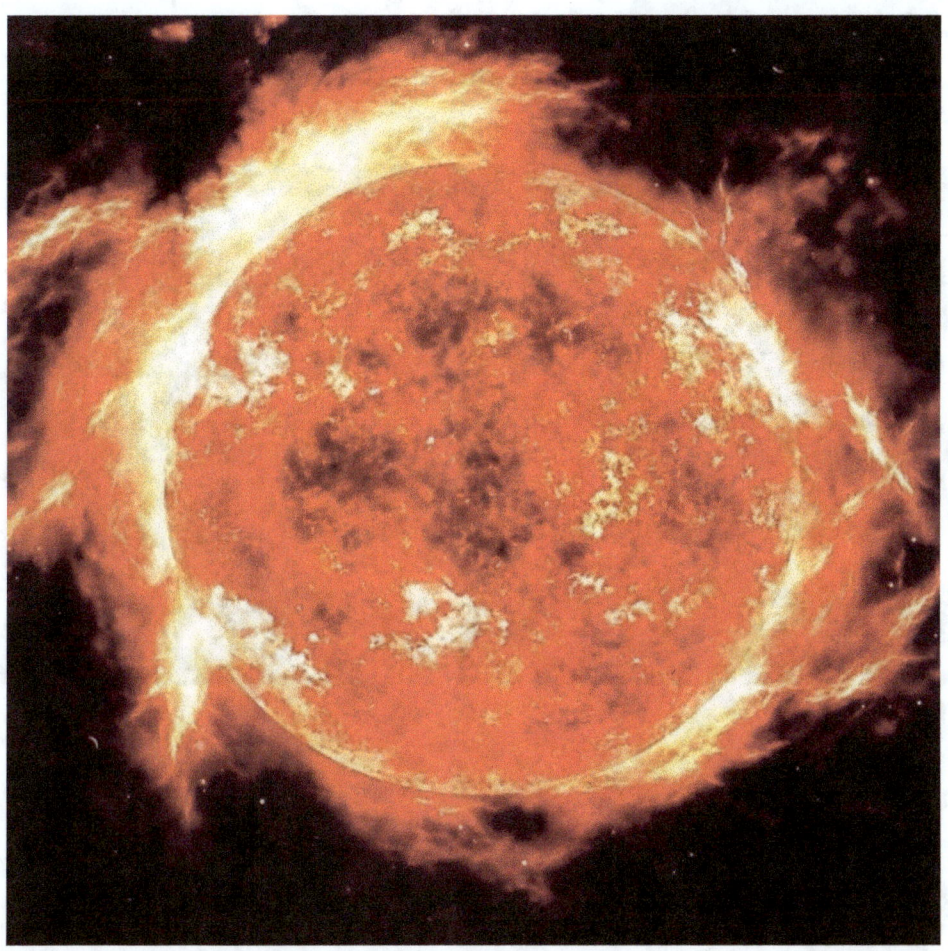

Understanding the structure and evolution of stars like UY Scuti is fundamental to studying the formation and evolution of galaxies and the universe as a whole. Additionally, red supergiant stars like this one play an important role in chemically enriching the interstellar medium, through the emission of heavy elements that are produced in their cores and propagated through space via stellar winds.

Finally, it is important to highlight that the observation and study of distant stars like UY Scuti are essential to broaden our knowledge about the universe and its complexity. Despite the technical difficulties involved, advances in astronomy have made it possible to obtain increasingly precise information about these objects, opening up new possibilities to explore the universe in which we live.

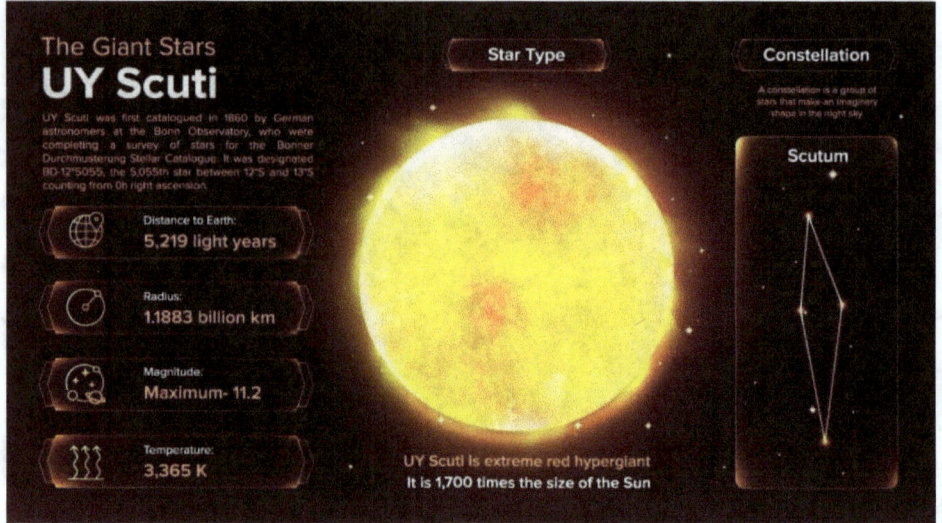

WOH G64

The star WOH G64 is a red supergiant located in the Large Magellanic Cloud, a satellite galaxy of the Milky Way. With an apparent magnitude of about 13, this star is very bright and can be seen with moderate-sized amateur telescopes.

One of the largest known stars, with an estimated radius of around 1,500 times the radius of the Sun, this red supergiant is also very massive, with an estimated mass of around 25 times the mass of the Sun.

Furthermore, WOH G64 is a very old star, with an estimated age of about 10 million years. The observation provides important information for understanding stellar evolution. Red supergiants like this star are late stages in the evolution of massive stars and provide clues about the evolution of massive stars. WOH G64 in particular is one of the most luminous stars known and can provide useful information about stellar evolution under extreme conditions.

Observations with telescopes in the visible and infrared spectrum reveal interesting features of the atmosphere of this star. For example, spectroscopic observations have revealed the presence of an expanded shell of gas around the star, called the circumstellar envelope. The presence of this envelope suggests that WOH G64 is undergoing an intense phase of mass loss, with the expulsion of large amounts of gas into its surroundings.

Other observations indicate that this star may be about to explode as a supernova. While it is not possible to accurately predict when this will happen, theoretical models suggest that it could happen in the near future, in astronomical terms.

The chemical composition of the star WOH G64 is an active topic of study among astronomers. However, spectral analysis of the star suggests that its atmosphere is rich in hydrogen and helium, as is common for stars. Additionally, trace amounts of heavier elements such as carbon, oxygen, and nitrogen were detected.

Spectroscopic observations of the star have also revealed the presence of some less common chemical elements in its atmosphere. For example, trace amounts of lithium, beryllium and boron were detected, which are normally difficult to detect in stars due to their low content. The presence of these elements suggests that WOH G64 may have undergone mixing and chemical enrichment processes during its stellar evolution.

Spectral analysis of the star suggests that it may be enriched in elements produced by advanced nuclear processes, such as the s-process and the r-process. These processes occur under extreme conditions, such as supernovae and neutron star collisions, and produce elements heavier than iron. The presence of these elements in WOH G64 may provide clues to the origin of these elements in high-mass stars.

RIGEL

Rigel's star is one of the brightest stars visible to the naked eye in the night sky. Located in the constellation Orion, it is a blue B-class supergiant star and has an apparent magnitude of about 0.18. Its position in the night sky makes it easily identifiable by both amateur and professional astronomers.

The Rigel star has an estimated mass of about 23 times the mass of the Sun and an estimated diameter of about 78 times the diameter of the Sun. It is a young star, estimated to be about 10 million years old. By comparison, the Sun is estimated to be about 4.6 billion years old. Rigel is located at a distance of about 860 light years from Earth.

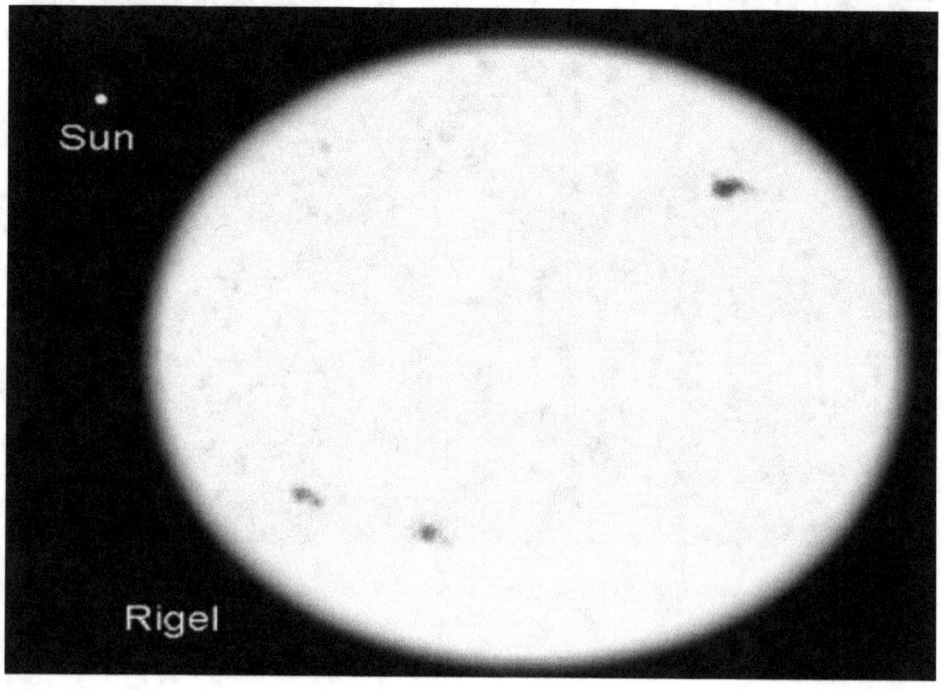

The bright blue color of the star Rigel is indicative of its relatively high surface temperature, estimated to be around 12,000 Kelvin. Rigel's high temperature means that it emits a lot of ultraviolet and visible radiation. This radiation is responsible for the star's luminosity and is also the source of energy for the ionization of gases in the surrounding interstellar medium.

Rigel is a variable star, which means that its luminosity varies slightly over time. The variation in the luminosity of the star is due to the pulsation of its surface, which can be observed as changes in the width of the spectral lines of its spectrum.

The star Rigel is also known to be a binary system, made up of a main star and a smaller companion. The exact nature of the companion is not well understood, but it is possible that it is a B or O minor star.

Due to its brilliant luminosity and location in the Orion constellation, the star Rigel has been the object of observation and study by astronomers for centuries. It is an important source of information on stellar evolution and stellar physics in general.

The chemical composition of the star Rigel is similar to that of other stars of its class. As a B-class blue supergiant star, it is made mostly of hydrogen and helium, like most stars. However, it also contains significant amounts of heavier elements such as carbon, nitrogen, oxygen, silicon, and iron.

The heavier elements are produced by nuclear fusion in the star's core, where temperatures and pressures are extremely high. During the life of a star like Rigel, it undergoes a series of nuclear reactions that produce these heavier elements. When the star reaches the end of its life, it can explode in a supernova, scattering these elements into space and enriching the galaxy with the elements that make up planets and other life forms.

Spectral analysis of the light emitted by the star Rigel can provide information about its chemical composition. Through spectroscopy techniques, astronomers can identify the spectral lines of different elements in your atmosphere and determine the relative abundance of those elements.

In general, the chemical composition of the star Rigel is very similar to that of other stars of its class, but the analysis of its spectral lines can provide important information about stellar evolution and the formation of elements in the universe.

The Rigel star has a very high rotation rate, rotating around its axis once every 10.4 Earth days. That is about 17 times faster than the speed of rotation of the Sun. Due to its high rotation speed, Rigel is a star flattened at the poles, with an equatorial diameter 50% greater than the polar diameter.

The orbit of this star is also of interest to astronomers. Rigel is a solitary star and is not part of a binary or multiple star system. However, it is located in the constellation Orion, which contains many bright young stars and moves relative to our solar system at a speed of about 24.4 km/s.

The orbit of the star Rigel around the galactic center of the Milky Way is estimated to be about 250 million years. This means that since Rigel formed, it has completed about 4 orbits around the galactic center. Rigel's position in the night sky is also constantly changing due to the star's own movement in space. Proper motion is the apparent change in the position of a star in the night sky relative to other background stars caused by the actual motion of the star in space.

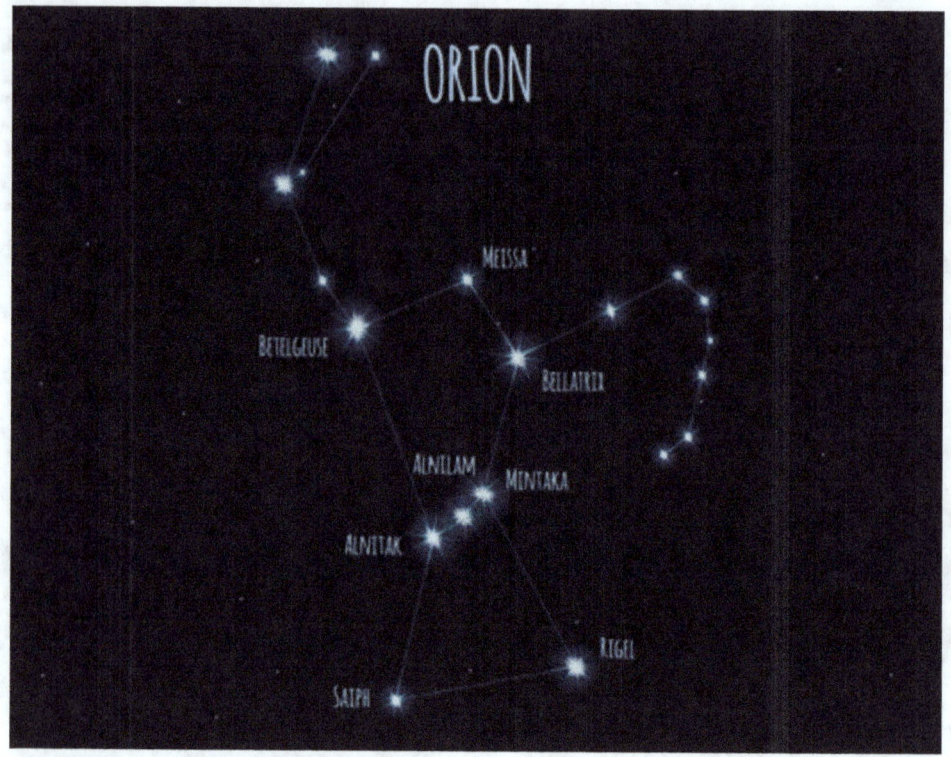

BLACK STARS

Black stars are a rare and intriguing astronomical phenomenon that has aroused the interest of the scientific community. Unlike conventional stars, black stars do not emit visible light and are therefore difficult to detect. In this chapter, we will discuss what black stars are, how they are formed, and what is their role in the universe.

What are the black stars? Black stars are extremely compact and dense stars, with such a mass that the force of gravity is able to prevent light from escaping from them. Because of this, they do not emit visible light and are virtually invisible to conventional telescopes. Their existence can only be detected through the gravitational effects they exert on other stars and nearby celestial objects.

These stars are formed from the explosion of massive stars, known as supernovae. During a supernova, the star explodes and the remaining core is compressed by an extremely strong gravitational force, forming a neutron star. If the mass of the neutron star is even higher, it can collapse further and form a black star.

These stars play a fundamental role in the universe, since they are in charge of maintaining the stability of galaxies. The gravitational pull of dark stars keeps stars and planets close to them in orbit, preventing them from escaping into intergalactic space. Furthermore, black stars may also play an important role in the production of cosmic rays and the formation of black holes.

A dark star need not have an event horizon and may or may not be a transition phase between a collapsing star and a singularity. A dark star is created when matter is compressed at a rate significantly less than the free fall velocity of a hypothetical particle falling towards the center of this star, due to the fact that quantum processes create vacuum polarization, which creates a form of degenerative pressure. preventing spacetime (and the particles trapped in it) from occupying the same space at the same time. This energy is theoretically unlimited, and if it builds up fast enough, it will prevent gravitational collapse from creating a singularity. This may imply a lower and lower collapse rate,

A black star with a radius slightly larger than the predicted event horizon for a black hole of equivalent mass will appear visibly very dim, because almost all of the light produced returns to the star. Any light that does escape will be severely affected by gravity, generating a redshift (also known as redshift) at that luminosity. It will appear almost exactly like a black hole.

Will have Hawking radiation[8], since virtual particles created in its neighborhood can still split, with one particle escaping and the other being trapped. Furthermore, it will create Planckian thermal radiation that resembles the expected equivalent Hawking radiation from a black hole.

The predicted interior of a black star will be composed of this strange state of space-time, with every depth length running inward, appearing the same as a black star of equivalent mass and radius without the shroud. Temperatures increase with depth toward the center.

NEUTRON STARS

Neutron stars are one of the most fascinating and enigmatic objects in the universe. They are compact remnants of massive stars that have run out of nuclear fuel and have gravitationally collapsed. Due to their incredible density, neutron stars have extreme physical properties, which make them the subject of great interest and study in astrophysics.

Neutron stars form from supernovae, which occur when a massive star uses up all its nuclear fuel and the gravitational pull of its core becomes untenable. At that moment, the core of the star collapses, forming an extremely dense sphere of matter, about 20 kilometers in diameter. This sphere is composed mainly of neutrons, which are subatomic particles with no electrical charge, and is surrounded by an atmosphere of electrons and protons.

The density of matter in neutron stars is so high that a teaspoon of their matter would weigh millions of tons on Earth. Also, neutron stars spin very quickly, with rotation speeds of up to hundreds of times per second. This rapid spin is a result of the principle of conservation of angular momentum, which causes the rate of rotation to increase as the star shrinks.

Neutron stars are detected through their emission of electromagnetic radiation, which can be observed in various bands of the electromagnetic spectrum, including X-rays, gamma rays, and radio waves. This radiation is produced by various physical processes that occur in neutron stars, such as rapid rotation, strong magnetic fields, and interaction with material in their surroundings.

One of the most intriguing properties of neutron stars is their extremely intense magnetic field, which may be billions of times stronger than Earth's magnetic field. This strong magnetic field creates a region of plasma around the star known as the magnetosphere, which interacts with the interstellar medium and can produce radio emissions.

In these systems, the stars orbit around a common center of mass and can interact gravitationally and through radiation emissions, producing complex and fascinating effects.

Neutron stars can also form binary systems with other stars, producing complex effects. The study of neutron stars is essential for understanding high-energy physics and the universe as a whole.

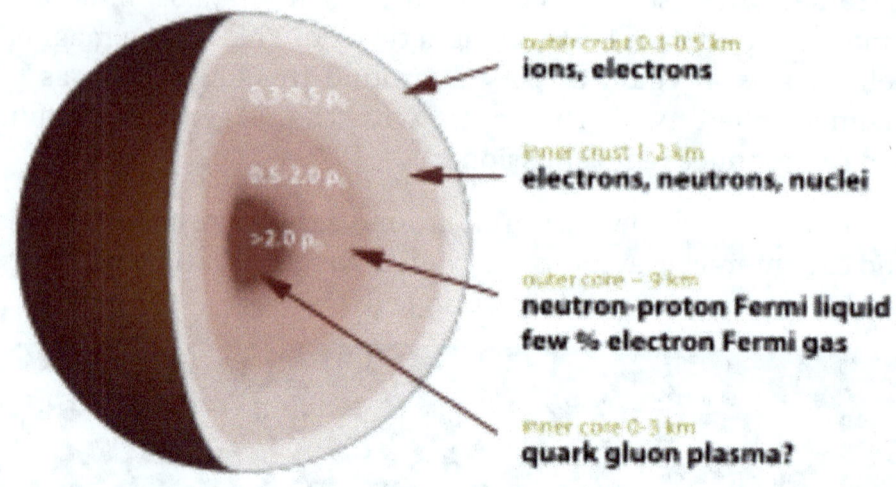

Structure of a neutron star

Pulsars are very small, very dense neutron stars. Pulsars can have a gravitational field up to a billion times that of Earth. They are probably remnants of collapsed stars or supernovae. As a star loses energy, its matter is compressed towards its center, becoming increasingly dense. The more the star's matter moves toward its center, the faster it spins.

They emit a constant flow of energy. This energy is concentrated in a current ofparticleselectromagneticwhich are issued frommagnetic polesof the star. As the star rotates, the beam of energy is scattered by thespace, like the packagelightof alighthouse. Only when the beam hits theLandis that we can detect pulsars through radio telescopes. The light emitted by pulsars in thevisible spectrumIt is so small that it is not possible to observe it fromnaked eye. Only radio telescopes can detect the strong energy they emit.

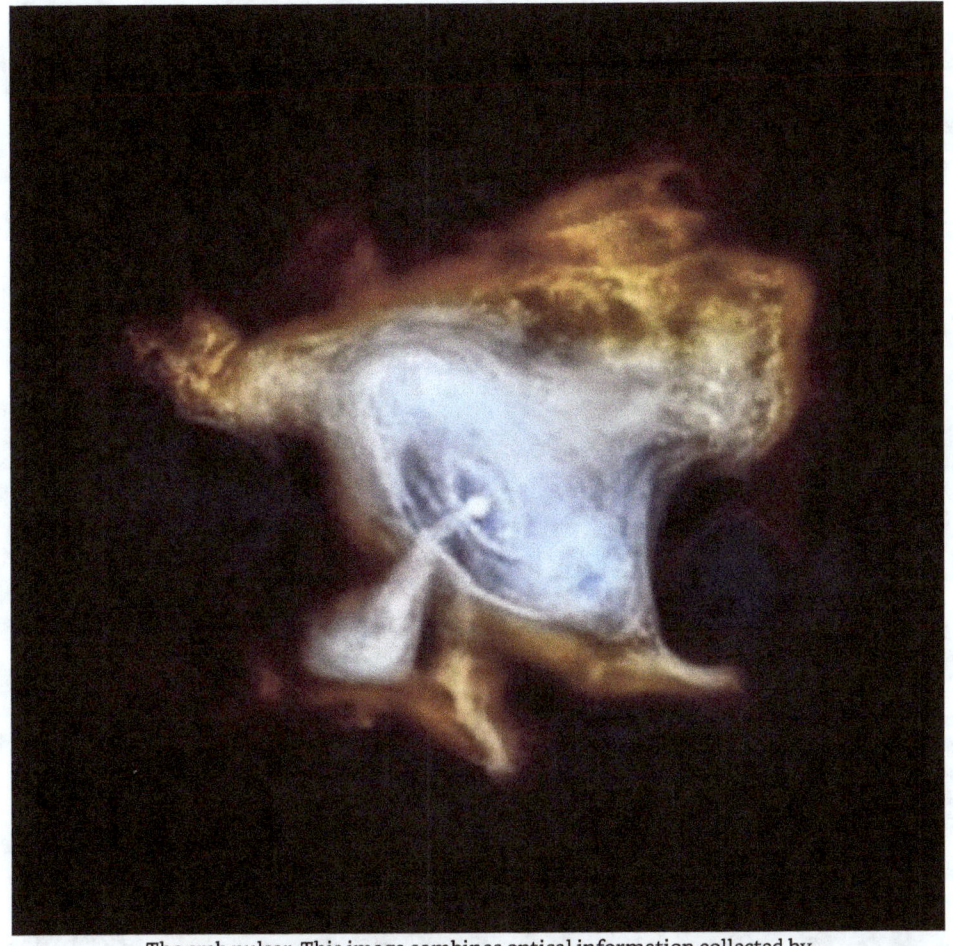

The crab pulsar. This image combines optical information collected by
Hubble (in red) and X-ray images from Chandra (in blue).

the pulsarREP 1913+16is a system orbited by neutron stars with
a maximum separation of a single radiussolarbetween them. It
is fast-moving, and observations indicate that the orbital period
of this system should decrease relatively quickly, given its strong
signal.gravitational wave; since 1975 the period has already
decreased by 10 seconds.

acceleration disc,in case of asuper newoccur in a binary system,
the companion supernova can sustain some damage to its surface
layers (and still continue to live), because each part of the binary

generates its own droplet-shaped domain of gravitational force, which coalesce into the form of a "8" forming aequipotential surface; call fromRoche lobe(all points have the same gravitational potential). A neutron star will form next to another neighboring star from the supernova. When the neighboring star becomes onered giant, fills the lobe, its gas will spiral towards the neutron star throughlagrange pointof the Lobe (unstable equilibrium point through which matter can be transferred). That gas that is sucked into the neutron star due to its rotation will form a thick disk around it; such a disk is calledaccretion.

The friction that exists between the gas layers in close orbits along the accretion disk leads to loss of angular momentum and spiraling downward toward the surface of the neutron star. The spiraling gas moves into the gravitational field of the neutron star, so its gravitational energy is converted to thermal energy inside the accretion disk.

In the inner part of the accretion disk, gravitational energy is released with greater intensity, reaching an average temperature of millions of degrees. An enormous source of energy is present in this region, where there is a large emission of radiation, such as ultraviolet rays and X-rays. The pressure on the neutron star can increase considerably if a relatively large amount of gas is transferred from the disk of neutron star accretion; in this way energy is accumulated and thus, eventually, gas is expelled from the neutron star, causing strong currents of gas in its orbit.

FINAL CONSIDERATIONS

At the end of this book on the stars of the universe, we can say that these celestial objects are true cosmic wonders. They are responsible for the creation of chemical elements, for the production of light and heat, as well as being one of the main elements that make up galaxies.

We have learned that stars can vary in size, temperature, color, and brightness, which can significantly affect their life cycle and eventual fate. Some stars end up exploding in supernovae, while others can become black holes or neutron stars.

Stars also play an important role in our very existence, as they are responsible for the light we see during the day, for warming our planet, and for providing essential elements for life, such as carbon and oxygen.

However, much remains to be discovered about the stars and the universe in which we live. As science advances, new technologies and research methods allow us to study stars and better understand their origin, evolution, and role in the cosmos.

In short, this book has shown us the magnitude and complexity of the stars in the universe and how essential they are to our understanding of the cosmos and our very existence.

BIBLIOGRAPHIC REFERENCES

Anglada-Escude, Guillem; et al. (August 2016). "A terrestrial planet candidate in a temperate orbit around Proxima Centauri". Nature. 536 (7617): 437-440. Bib code: 2016Natur.536..437A. doi:10.1038/nature19106

Baker, J.; Bizarro, M.; Wittig, N.; Connelly, J.; Hack, H. (2005). "Early planetesimal merger from an age of 4.5662 Gyr for differentiated meteorites". Nature. 436: 1127–1131. doi:10.1038/nature03882

Barcelo, C.; Liberati, S.; Sonego, S.; Visser, M. (2008). "Fate of gravitational collapse in semiclassical gravity". Physical Review D 77:044032. doi:10.1103/PhysRevD.77.044032. (in English)

Bessa Soares (February 9, 2011). The Sun is a perfect sphere. More technology. Accessed June 30, 2021

Bonano, A.; Schlattl, H.; Paterno, L. (2008). "The age of the Sun and the relativistic corrections in the EOS". Astronomy and Astrophysics. 390: 1115–1118. doi:10.1051/0004-6361:20020749

Camenzind, Max (February 24, 2007). Compact Objects in Astrophysics: White Dwarfs, Neutron Stars, and Black Holes Springer Science & Business Media. P. 269. ISBN 978-3-540-49912-1

Dearborn, David SP (2016). "Evolutionary Clues for Betelgeuse". The Astrophysical Journal. 819. 7 pages. Bib code:2016ApJ...819....7D. arXiv:1406.3143v2. doi:10.3847/0004-637X/819/1/7

DeWarf, LE; Datin, KM; Guinan, E.F. (October 2010). "X-ray,

FUV and UV observations of α Centauri B: determination of the long-term magnetic activity cycle and period of rotation". The Astrophysical Journal. 722(1): 343-357. Bib Code:2010ApJ...722..343D. doi:10.1088/0004-637X/722/1/343

Dolan, Michelle M.; Mathews, Grant J.; Lam, Doan Duc; Lan, Nguyen Quynh; Herczeg, Gregory J.; dos Anjos, Sandra. Evolution of Stars in Binary Systems (PDF). Institute of Astronomy, Geophysics and Atmospheric Sciences: University of São Paulo.

Edward F. Guinan; Richard J Wasatonic; Thomas J. Calderwood (December 8, 2019). "ATel #13341: The fainting of the nearby red supergiant Betelgeuse". The astronomer's telegram. Consulted on January 11, 2023

ESO: Highest resolution image of Eta Carinae obtained to date incl. Images and animation
The study shows that the Sun is the most perfect sphere in nature. www.apolo11.com. Accessed June 30, 2021

G. Wallerstein; I. Iben son; P. Parker; AM Boesgaard; G.M. Hale; Champagne AE; , CA Barnes; F. KM-dppeler; VV Smith; RD Hoffman; special effects
Times; C.Sneden; RN Boyd; B. S. Meyer; D. L. Lambert (1999).

See GCVS=Eta+Car». General Catalog of Variable Stars @ Sternberg Astronomical Institute, Moscow, Russia. Consulted on November 12, 2022

Glendenning, Norman K. (2012). Compact Stars: Nuclear Physics, Particle Physics, and General Relativity Illustrated Ed. [SL]: Springer Science & Business Media. P. 1. ISBN 978-1-4684-0491-3 Page Extract

Godier, S.; Rozelot, J.-P. (2000). Solar flattening and its relationship with the structure of the tacocline and the solar subsoil (PDF). Astronomy and Astrophysics. 355: 365–374. Bib Code:2000A&A...355..365G

Haensel, Pawel; Potekhin, Alexander Y.; Yakovlev, Dmitry G. (2007). Neutron stars. [SL]: Springer. ISBN 0-387-33543-9

Ham, W. T. Jr.; Muller, HA; Ruffolo, JJ Jr.; Guerry, D. III, (1980). «Solar retinopathy as a function of wavelength: its significance for protection

Glasses". In: Williams, TP; Baker, BN The effects of constant light on visual processes. [Sl]: Full Press. pages. 319–346. ISBN: 0306403285

Harper, G.M.; et al. (July 2017). "An updated 2017 astrometric solution for Betelgeuse". The Astronomical Journal. 154 (1): article 11, 6 pp. Bib Code: 2017AJ....154...11H. doi:10.3847/1538-3881/aa6ff9

Helerbrock, Raphael. «What is a neutron star?. Brazil School. What is Physics?. Omnia network. Retrieved December 21, 2022

Hitchcock, R. Timothy; Patterson, Patterson (1995). Radiofrequency Electromagnetic Energies and ELF: A Handbook for Healthcare Professionals. [EN]: John Wiley and Sons. P. 218. ISBN: 9780471284543

Howard RA; Moses JD; Socker DG; Dere KP; Cook JW (2002). "Sun Earth Connection Coronal and Heliospheric Research (SECCHI)". Solar Variability and Solar Physics Missions Advances in Space Research. 29(12): 2017–2026

Keenan, Philip C.; McNeil, Raymond C. (October 1989). "The Perkins Catalog of Revised MK Types for the Coolest Stars". Astrophysical Journal Supplement Series. 71:245-266. Bib code:1989ApJS...71..245K. doi:10.1086/191373

Kervella, P.; Mignard, F.; Merand, A.; Thévenin, F. (October 2016). "Close stellar conjunctions of α Centauri A and B until 2050. A star mK = 7.8 may enter the Einstein ring of α Cen A in 2028." Astronomy and Astrophysics. 594: A107, 15.

Kiziltan, Bulent (2011). Fundamentals reassessed: on the evolution, ages and masses of neutron stars. [Sl]: Universal Editorials. ISBN 1-61233-765-1

Lodders, K. (2003). "Abundances of the Solar System and Condensation Temperatures of the Elements". Astrophysical Journal. 591 (2): 1220. doi:10.1086/375492

Miglio, A.; Montalbán, J. (October 2005). "Constraining fundamental stellar parameters using seismology. Application to α Centauri AB". Astronomy and Astrophysics. 441(2):615629. Bib Code:2005A&A...441..615M. doi:10.1051/0004-6361:20052988

Montarges, M.; Kervella, P.; Perrin, G.; Chiavasa, A.; Le Bouquin, J.-B.; Auriere, M.; Lopez Ariste, A.; Mathias, P.; Ridgway, ST; Lacour, S.; Haubois, X.; Berger, J.-P. (2016). "The near circumstellar environment of Betelgeuse. IV.

VLTI/PIONIER interferometric monitoring of the photosphere". Astronomy and Astrophysics. 588:A130. Bib Code:2016A&A...588A.130M. arXiv:1602.05108. doi:10.1051/0004-6361/201527028

NASA satellites capture the start of a new solar cycle. PhysOrg (Science/Physics News). January 4, 2008. Accessed July 10, 2022. POT. «The RXTE X-ray light curve of Eta Carinae

O'Gorman, E.; et al. (August 2015). "Time evolution of the size and temperature of Betelgeuse's extended atmosphere". Astronomy and Astrophysics. 580: A101, 11 pp. Bib Code:2015A&A...580A.101O. doi:10.1051/0004-6361/201526136

Orel, Thierry (August 2018). "Review of the chemical composition of α Centauri AB". Astronomy and Astrophysics. 615: A172, 22.

Paardekooper, S.-J.; Leinhardt, ZM (March 2010). "Planetesimal collisions in binary systems". Monthly Notices of the Royal Astronomical Society: Letters. 403(1): L64-L68.

Phillips, 1995, pp. 78–79 Pesquisa Fapesp Magazine (March 8, 2012). «Fapesp research magazine: Eta carinae, beyond the eclipse Robrade, J.; Schmitt, JHMM; Favata, F. (October 2005). "X-rays from α Centauri - The dimming of the solar twin". Astronomy and Astrophysics. 442(1): 315-321. Bib Code:2005A&A...442..315R. doi:10.1051/0004-6361:20053314

Samus, NN; Kazarovets, EV; Durlevich, OV; Kireeva, NN; Pastukhova, IN (January 2009). "VizieR Online Data Catalog: General Catalog of Variable Stars (Samus+, 2007-2017)". VizieR Online Data Catalog: B/gcvs. Bib Code:2009yCat....102025S

Schutz, Bernard F. (2003). Gravity from zero. [SL]: Cambridge University Press. pages. 98–99. ISBN 9780521455060

Seidelmann; et al. (2000). Report of the IAU/IAG Working Group on Cartographic Coordinates and Elements of Rotation of Planets and Satellites: 2000». Retrieved March 22, 2006

Result of the basic query SIMBAD». SIMBAD. Consulted on January 9, 2023
Sol. Dictionary of Aulete. Retrieved April 14, 2010. Archived from the original on July 6, 2022.

The vital statistics of the sun. Stanford Solar Center. Accessed on July 29, 2008, citing Eddy, J. (1979). A new sun: Skylab's solar results. [EN]: NASA. P. 37. NASASP-402

Visser, Matt; Barcelo, Carlos; Liberati, Stefano; Sonego, Sebastiano (2009) "Small, dark and heavy: But is it a black hole?", Bibcode: 2009arXiv0902.0346V

Woolfson, M. (2000). "The origin and evolution of the solar system". Astronomy and Geophysics. 41. 1.12 pages. doi:10.1046/j.1468-4004.2000.00012.x
Zeilik, MA; Gregorio, SA (1998). Introduction to astronomy and astrophysics 4th ed. [Sl]: Saunders College Publishing. P. 322. ISBN 0030062284

Zhang, Bing; Xu, RX; Qiao, G.J. (2000). "Nature and nurture: a model for soft gamma-ray repeaters". The Astrophysical Journal. 545(2): 127–129. Bib code:2000ApJ...545L.127Z. arXiv:astro-ph/0010225. doi:10.1086/317889. Consulted on September 22, 2021

Zhao, lily; Fisher, Debra A.; Brewer, John; Giguere, Matt; Rojas-Ayala, Barbara (January 2018). "Detectability of planets in the Alpha Centauri system". The Astronomical Journal. 155 (1): article 24, 12.

---

[1] Inastronomy, perihelion (or perihelion), which comes from peri (around, near) and helium (Sun), is the point oforbitof a body, eitherplanet,dwarf planet,asteroideitherkite, which is closer toSun. When a body is at perihelion, it has the greatestspeedintranslationof its entire orbit. When the body in question is orbiting any other celestial object than the Sun, the generic name is used.periastromto identify that point.

[2] aphelionis the point oforbitin whichplanetor oneminor body of the solar systemis further fromSun. When it is an object orbiting a star other than the Sun, this point is calledapostrophe. The orbits of all the planets are alwayselliptical, always having a more distant point (aphelion) and a closer point (perihelion).

[3] unitBased onInternational System of Units(YES) for greatnessthermodynamic temperature. The kelvin is the fraction $1/273.16$ of the thermodynamic temperature of thetriple pointfrom thewater, that is, it is defined such that the triple point of water is exactly 273.16 K

[4] Technique used to estimate the age of objects and events.astrophysicists. This technique uses the abundance of radioactive nuclei, such asuraniumIsthorium, similar to usingCarbon-14incarbon dating.

[5] Determining the age of an object from substances.radioactivecontained therein and the products of theradioactive decay

[6] In astronomy, stellar parallax is used to measure the distance to stars using the Earth's motion in its orbit. It is the angle formed by the rays that start from the center of a star and will have, one at the center of the Earth, another at the point where the observer is.

[7] Nucleosynthesis is the process of creating new atomic nuclei from pre-existing nuclei to generate the rest of the elements on the periodic table.

[8] This radiation was predicted from theoretical considerations of both thegeneral theory of relativityhow much ofclassical thermodynamics. The original line of

reasoning was drawn by an Israeli scientist namedjacob bekenstein, who had suggested that black holes might have aentropywell-defined, which, in turn, would suggest that they also have atemperatureequally well defined. In light of this prediction, Hawking radiation is sometimes called Bekestein-Hawking radiation.

ABOUT THE AUTHOR

José Ruiz Watzeck

Journalist, Writer, Author, Geographer, Mathematician, Professor, Neuropsychopedagogue, Specialist in Higher Education Teaching, Postgraduate in Auditing, Management and Environmental Licensing, Postgraduate in Geoprocessing and Georeferencing, Pedagogue.

BOOKS BY THIS AUTHOR

The History Of Astronomy: From Prehistory To The 20Th Century

Astronomy is the oldest of sciences. Archaeological discoveries have provided evidence of astronomical observations among prehistoric peoples. Since ancient times, the sky has been used as a map, calendar, and clock. The oldest astronomical records date from approximately 3000 BC and are due to the Chinese, Babylonians, Assyrians and Egyptians. At that time, the stars were studied for practical purposes, such as measuring the passage of time (calendars), predicting the best time for planting and harvesting, or for purposes more related to astrology, such as making predictions about the future, who believed that the sky gods had the power of harvest, rain and even life.

By studying megalithic sites such as those at Callanish in Scotland, the Stonehenge circle in England, dating from 2500 to 1700 BC. C., and the Carnac alignments in Brittany, astronomers and archaeologists have concluded that the alignments and circles served as landmarks indicating references. and important points on the horizon, such as the extreme positions of the rising and setting of the Sun and the Moon, throughout the year. These megalithic monuments are authentic observatories for predicting eclipses in the Stone Age.

At Stonehenge, each stone weighs an average of 26 tons. and the main avenue that runs from the center of the monument points to the place where the sun rises on the longest day of summer. In this structure, some stones are aligned with sunrise and sunset in early summer and winter. The Maya in Central America also had knowledge of the calendar and celestial phenomena, and the Polynesians learned to navigate through celestial observations.

www.ingramcontent.com/pod-product-compliance
Lightning Source LLC
Chambersburg PA
CBHW070354220526
45467CB00001B/375